P9-EEJ-887

Charles Darwin and the Problem of Creation

Neal C. Gillespie

Charles Darwin and the Problem of Creation

The University of Chicago Press

Chicago and London

The University of Chicago Press, Chicago 60637
The University of Chicago Press, Ltd., London

Printed in the United States of America
83 82 81 80 79 5 4 3 2 1

Library of Congress Cataloging in Publication Data

Gillespie, Neal C 1932–
Charles Darwin and the problem of creation.

Bibliography: p.
Includes index.
1. Darwin, Charles Robert, 1809–1882.
2. Life—Origin. 3. Naturalists—England—
Biography. I. Title.
QH31.D2G55 575'.0092'4 79–11231
ISBN 0–226–29374–2

Neal C. Gillespie is professor of history at Georgia State University. He is the author of *The Collapse of Orthodoxy: The Intellectual Ordeal of George Frederick Holmes.*

For my parents

He giveth to the Beast his food, and to the young Ravens when they cry.

Psalm 147:9

He remembered the sense of loss and disgust and horror when he saw it: it swam upward wriggling heavily in a flail of heavy dying protest, through a thickened murk of greenish water, and he saw that to its brain was fastened some blind horror of the sea, a foul snake-like shape a foot or more in length, a headless, brainless mouth, a blind suck and sea-crawl, a mindless abomination, glued implacably, fastened in fatal suck in one small rim of bloody foam against the brain-cage of the great dying fish.

Thomas Wolfe, *Of Time and the River*

Contents

Preface

The true introduction to this book is in the first two chapters, so there is no need for a lengthy preface. Nonetheless, I might use a prefatory opportunity to indicate to the reader how I hope this book relates to current work on the career of Charles Darwin.

This book is predominantly about Charles Darwin and the world of naturalists in which he lived, but it is not about his discovery of evolution by means of natural selection. It is not an assessment of the evidence for and against that view as the question was argued by Darwin and his contemporaries; nor is it the story of the writing and subsequent career of the *Origin of Species.* These topics are important, and there are many excellent works dealing with them, and there will, I am sure, be many more. The reader who is familiar with the recent growth of Darwin studies will know my debt to them; nor are the books and articles discussed and cited herein any measure of that debt.[1]

In the same way, I have not dealt here with the derivation and content of Darwin's several other scientific theories as such, nor with the question of the psychological, social, and other sources of his ideas. These issues, like natural selection, are important, and they, too, have become objects of study by many fine scholars. But the focus of this book is elsewhere.

On reading the *Origin of Species,* I, like many others, became curious about why Darwin spent so much time attacking the idea of divine creation. This interest led me to notice that the book had a surprising amount of positive theological content. What was the significance of this, and what was the relationship, if any, between it and Darwin's scientific doctrines? How serious was Darwin's interest in theological questions? What was the place of that concern in his more general thought about science? Further study seemed to show that he was quite seriously (if not profoundly) concerned with religious questions, and that his concern was involved, in positive and negative ways and to a degree frequently overlooked, in his idea of good science. Obviously, Darwin thought the *Origin* was good science. Yet behind the arguments of that volume lay prior beliefs about knowledge—particularly scientific knowledge—about nature and even about God. How were these ideas connected and their apparent conflicts resolved, or not, in Darwin's thought? Did these ideas reflect a more basic antagonism within the natural history of his time than the famous conflict over evolution? In what ways *was* creation a problem for Darwin?

More than this would anticipate too much of what follows. But perhaps enough has been said to indicate the particular slant of this study. I am concerned, then, with what might be called the existential dimension of Darwin's work: his attitude toward human knowledge and its possibilities, toward religious faith and its place in human experience. The inquiry is made somewhat difficult by Darwin's well-known reticence on such subjects. Often, one must catch him unaware, as it were, and make what one can of obscure and ambiguous materials. But this, to follow the trails where they lead and to venture into the wilderness when the trails run out, is, after all, the historian's task.

I would not leave the impression that these questions have never before been asked. I do think, however, that they have seldom received the emphasis they deserve. There is, furthermore, a tendency in some of the recent Darwinian studies to "modernize" the Sage of Down by treating his vision of nature as thoroughly naturalistic and the man as completely secular. I hope to show in this book why the first of these "modernizations" is both understandable and deeply unhistorical. At the same time, however, it is not entirely wrong. The theory of evolution that Darwin gave to the world was "naturalistic," or, as I shall call it, positivistic. But that was not Darwin's theory. One of the most striking features of the natural history of Darwin's time was the emergence of a "naturalistic" biology. Such a biology is ours, but it was not Darwin's—or was only partly so. As for the second "modernization": that the theological language in Darwin's writing should not be dismissed as merely poetic or rhetorical or as some sort of elaborate deception without examination of the case for its seriousness is one of the major propositions of this book. Obviously, I find the case for a serious reading both compelling and broadening of our understanding of Darwin and his time. Seen in this way, Darwin was his own scientific world writ small; and it is in the paradoxical relationship between his science and his theology that we can find a paradigm of change whereby one world view passed into another. The reader, of course, must judge for himself.

A further word of caution: I have used several terms, particularly "creationism," "positivism," and "biblicism" in special senses which are defined in the text. The reader is asked to remember these usages as he reads.

In addition to the debt which I owe to other scholars, I wish to thank the College of Arts and Sciences of Georgia State University for the released time from other duties which aided in the completion of this study, my former chairman Joseph O. Baylen and my other colleagues in the Department of History for their interest, and the staffs of the William Russell Pullen Library of Georgia State University and the Robert W. Woodruff Library of Emory University for their assistance. As always, my wife Sherrie was there when she was needed.

1 Positivism and Creationism: Two Epistemes

"All science," said Emerson, "has one aim, namely, to find a theory of nature." He perhaps should have said that all science rests on a theory of nature and makes no sense apart from it. This book might have been subtitled "epistemic and paradigmatic tensions in the *Origins of Species.*" The reader would have been given a better clue to my purpose thereby, but misled at the same time. In analyzing some of the central intellectual issues that underlay the problem of special creation in nineteenth-century natural history, I want to draw on the insights of Michel Foucault and Thomas S. Kuhn into the composition and dynamics of communal thought as presented in their influential books *Les mots et les choses* and *The Structure of Scientific Revolutions.* My desire is to use this perspective, however, not as dogma, but as an organizing principle. I make no claim to embrace all of Foucault's profound, if sometimes extravagant, ideas about "epistemes,"—i.e., the communal presuppositions about knowledge and its nature and limits—nor to follow their subsequent mutations.[1] Nor do I seek to test Kuhn's model of changing paradigms in scientific thought in any narrow sense. Rather, I merely wish to employ what I take to be a useful concept and, at the risk of jargonizing, a particular terminology which has become familiar to historians—perhaps too familiar—and which I hope can contribute to a fuller understanding of Charles Darwin and the particular and complex intellectual world in which he worked: a world superficially similar to that of twentieth-century biological science, but one different in many important ways.

In the foreword to the English edition of *Les mots et les choses,* Foucault invited the reader "to make what [use] he will of the book he has been kind enough to read."[2] I propose to do this. I am sure that I could not have written this book as it now exists within the prescriptions laid down by Foucault. I would then have been concerned, not with the consequences of a positive perspective on the field of natural history and on the origin of evolutionary theory, but with the "rules" within that "discourse" and their relation to other contemporary areas of thought. This book, then, is in no sense a work of "archaeology."[3] My use of Foucault's concept of an episteme will be rather free, an illegitimate offspring perhaps, but one consciously inspired by and indebted to his fascinating book. Needless to say, he is totally innocent of anything in the present volume other than this initial organizing principle.

What does Foucault mean by an episteme? An episteme is the "historical *a priori*" that

> in a given period, delimits in the totality of experience a field of knowledge, defines the mode of being of the objects that appear in that field, provides man's everyday perception with theoretical powers, and defines the conditions in which he can sustain a discourse about things that is recognized to be true.

He writes, further:

> What I am attempting to bring to light is the epistemological field, the *episteme* in which knowledge, envisioned apart from all criteria having reference to its rational value or to its objective forms, grounds its positivity and thereby manifests a history which is not that of its growing perfection, but rather that of its conditions of possibility. . . .[4]

To investigate these "conditions of possibility" in the science of Darwin and his contemporaries is one object in this book.

What does Kuhn mean by paradigms? They are "universally recognized scientific achievements, that for a time provide model problems and solutions to a community of practitioners." A paradigm is a synthesis of sufficient scientific merit to draw practitioners away from rival theories and which functions as a source of future methods, questions, and problems. When anomalies, which inevitably arise out of research, reach a certain intensity, thereby initiating a "crisis," a change of paradigm occurs. This Kuhn defines as "a reconstruction of the field from new fundamentals, a reconstruction that changes some of the field's most elementary theoretical generalizations as well as many of its paradigm methods and applications." There is some temporary overlap between paradigms, but when the change is completed, "the profession will have changed its view of the field, its methods, and its goals." Kuhn's use of the term paradigm is loose. Sometimes it means a particular theory of narrow application; at other times, he uses it to signify a broader, general perspective on science and nature.[5] In this latter sense, Foucault's term episteme seems to me more apt.

Foucault deals with the "positive unconscious of knowledge: a level that eludes the consciousness of the scientist and yet is part of scientific discourse," the "epistemological space specific to a particular period."[6] This is analogous to and yet broader than Kuhn's usual use of "paradigm." The latter governs a specific field of science, its questions, anticipated answers, assumptions, and techniques of inquiry. An episteme, on the other hand, involves the idea of science itself: what constitutes good science; its structure, goals, and limits; and how science is even possible. These ideas are articulated, i.e., "conscious," in a few, but by and large they are "unconscious" in scientists. They are learned through work and training and become axiomatic in the understanding of both science and nature. Both of these concepts, Kuhn's and Foucault's, have more precision than the familiar "world view" which blends together virtually everything, or the elastic "ideology" which alternately contracts or sprawls over the entire intellectual map, according to the interest of its user. I shall use them, therefore, in the narrow and more general senses indicated. Paradigms guide research and epistemes provide them with the logic, metaphysics, and epistemology that make scientific work possible.[7]

Several historians have expressed skepticism as to whether the rise of evolutionary biology fits Kuhn's model.[8] In a close sense, it may not. In any event, it would be distracting to address the question here. What I am suggesting is the utility of such analysis in a looser sense: not as rigid formula but as a means of detecting changes in attitudes about science in nineteenth-century Anglo-American natural history, and of exploring the consequences of those changes. It is an application of what Darwin liked to call the "as if" approach to inquiry. Let us suppose that something like a shift in episteme or paradigm did occur, and see where it leads.

"In any given culture and at any given moment," Foucault wrote in *Les mots et les choses,* "there is always only one *episteme* that defines the conditions of possibility of all knowledge, whether expressed in a theory or silently invested in a practice."[9] This observation, which might otherwise serve as a motto for this book, immediately presents difficulties: Only *one* episteme? *All* knowledge? As shall be seen, a large part of the conflict in Darwin's era arose from the fact that there were, in effect, not one but two major epistemes in natural history invoking different standards of scientific knowledge and influencing in multitudinous ways the practice of naturalists as well as their theories about nature. I shall call them positivism and creationism.[10] The positivist limited scientific knowledge, which he saw as the only valid form of knowledge, to the laws of nature and to processes involving "secondary," or natural, causes exclusively. The creationist, on the other hand, saw the world and everything in it as being the result of direct or indirect divine activity. His science was inseparable from his theology. His epistemology was closely geared to a metaphysics, and in metaphysics he tended to be an "idealist". To comprehend nature fully, for a scientist of this persuasion, was to understand the workings of the mind of the Creator. This emphasis on mind, purpose, and design in nature is what I shall mean by "idealism."[11]

I hasten to point out the artificial nature of these epistemes. They are historian's constructs and are justified, if at all, by their heuristic value. They correspond, closely I think, to historical realities, but necessarily involve a schematic analysis. I am not suggesting that all naturalists in any given period should be lumped together in either category. No man mirrored them exactly, and most naturalists partook of both in different degrees over the period under consideration. No man instantly stepped from one episteme into the other as through a rent in time. Instead, the new conception of science and nature developed within the practice of the old and grew out of the needs and internal tensions that developed in that practice. Its birth was a gradual process varying from naturalist to naturalist and taking more than a generation to be accomplished.

Foucault's acuteness is revealed in the fact that without turning one's attention to the "silent practice" of naturalists it would not be possible to comprehend how the conflict between these two disparate views of science and nature arose or why positivism triumphed over its rival. But his schematic rigidity introduces a second problem that I also hope to help to clarify.

In *Les mots et les choses,* Foucault stressed a disjunction between eighteenth and nineteenth-century life science so absolute as to lead him to deny that the ideas we know as biology, life, or evolution even existed conceptually until after 1800. These paradoxes, which are more defensible than they seem at first glance, are the consequence of his epistemic discontinuities. Escaping these apparent perversities forces him into dubious verbal distinctions that obscure real and valid historical continuities.[12] Some followers of Kuhn and commentators on his work (if not Kuhn himself) have also emphasized the problem of discontinuity in the historical sequence of scientific theory.[13] And this, in turn, has raised the problem of "incommensurability," i.e., a subjective relativism in science and in its standards of judgment. It is not my purpose to attempt to resolve either of these difficult philosophical problems. I would only suggest that the use of an individual or collective biographical perspective may provide some correctives to the more extreme formulations of discontinuity and relativity that have attended discussion of the "incommensurability" problem.

One source of difficulty, I think, lies in abstraction.[14] It seems to have been assumed by some writers that, because sequential epistemes and paradigms may be stated as logically autonomous, internally consistent, and self-contained models of discourse, therefore they must be envisioned as radically discontinuous historically. So conceived, it is not strange that how men could move from one episteme to another, or how there could be independent judgment of the superiority of one paradigm over another, should become serious problems. While it would be absurd of me to presume to tell philosophers of science how to conduct their discipline, and with a full sense of the difficulty of these questions, I cannot help but think that some of the well-known paradoxes in this discussion are, to some extent, the result of abstracting the problem. Thought, severed from its object, as Bacon wrote, turns inward and spins cobwebs out of its own substance.[15]

The place of nonrational elements in science is one of the most important topics in the current reevaluation of the place of science in our world, and it merits our full attention. But we must base our inquiry on as full a knowledge as can be acquired of the actual process of knowing as it has actually occurred. Despite the use of historical examples, many discussions imply a rigid and reductive intellectual determinism that, if taken literally, would truly make it hard to see how men could ever change their minds. What is left out, of course, is the history, the personal experience through time. Abstract examinations tend to slight that area in which truth is most likely to be found: in the encounter of intelligence with the world. To attempt to resolve these problems by abstract analysis alone is to court paradox. But if, instead of merely seeing changes between self-existent and self-justifying constellations of ideas as the reality to be understood, we see that reality as one of men changing their ideas—a reality not of thought mysteriously changing, but of men thinking—the problem seems to me to be lessened. For, in history, epistemes and paradigms are not discontinuous. They are dialectically related through human experience. Men are not in one episteme and then in another with no independent judging activity; rather, they move from one mode of discourse or from one paradigm to another as the change meets the needs of their work. Ideas are changed and judgments validated in the course of working out a new way of doing science. Scientists, I suggest, at least initially, do not believe one episteme is superior to another because they are in it; rather, they are in it *because* they believe it is superior, and this belief is a product of their experience.

Again, I would not be taken to say more than I intend. Philosophical analysis of scientific thought, important as it is, is only part of our understanding of science. The historical reality of science is a complex of variables, some of which are left out of abstract analysis, yet which make a difference in the result. The process whereby many of Darwin's generation gradually lost the ability to see the world as the creationists saw it is a study in the development of the incommensurability of scientific ideas. But the disability to understand the ideas of another episteme is not automatically acquired with a new system; it grows with practice. Incommensurability is a function of time and the developing conventions of scientific work. In its classic form it is a characteristic of the second generation. The first generation displays a defensiveness that shows all too well that the old ideas still make sense to them.

A second, and related, possible error, I think, has been to identify the recent insistence on the historicity of science—that is to say, the understanding of science as a historical and cultural product—with a radical subjectivity of judgment so that scientific opinion is dissolved into a relativity of time and place. Surely this is to confuse two very different things. To say that scientific judgments are not objective—by this I mean *cannot* be objective, for no one denies that scientists any more than historians may be influenced by outside factors—but rather follow the desires and assumptions of scientists is to imply that the processes of objectivity (evidence, tests and experiments, logic, and so forth) have nothing to do with shaping those desires and assumptions. On the contrary, it would seem that these things are closely interrelated and that interrelation may best be seen on the workaday level of science where we may study the dialectical relation between the social inheritance and novel experience. Biography, as in the case at hand, can be an important guide in understanding this process of change.[16] Further, to say that the ideas of scientists are subjective or ideological because of their being influenced by a given social matrix, and to conclude, therefore, that they are not true—presumably because they do not result from some quest of pure reason untouched by the world—is surely a gross instance of the genetic fallacy. The source of an idea is irrelevant, in the strict logical sense, to its success in a scientific system or in any other. The procedures of proof in any knowledge system are logically independent of the circumstances of the origin of the ideas involved.[17] We need to know such things in order to fully understand science as a historical and social entity, but such knowledge, while it may caution one about scientific theories, cannot determine their truth. Historical or sociological judgments *about* science from within those perspectives are not judgments *in* science. A realization of the historicity of science should give us a greater comprehension of science, but should not lead to its abolition.

Although, in discussing the incommensurability problem, Kuhn likens paradigm change to a conversion experience, he does not believe it is necessarily instantaneous—as the frequently drawn analogy with religious conversion would suggest—nor does he seem unwilling to allow the evident role of facts, argument, and ratiocination in inducing change. Nonetheless, he may exaggerate the need for changing one's view "all at once."[18] As we shall see, Darwin and his colleagues were strikingly capable of impressive mental ambivalence over considerable periods of time.

Foucault, for his part, has protested against the charge that he ignores the problem of change or, worse, that his view of history makes a knowledge of change impossible. "My main concern," he says, "has been with changes," sudden and dramatic changes. But his defense shows that it is change after it has occurred that interests him. This, in a real sense, is not the study of change at all, but of two sequential static conditions examined comparatively. It is the contemplation of discontinuity that intrigues him, of disjunction and of contrasts. He sets aside the problem of causality as, for now, beyond solution. Hence, while he compares epistemes, he cannot traverse the space between them. He might well ask historians how "knowledge succeeds in engendering knowledge, ideas in transforming themselves and actively modifying one another." In these terms, it is doubtful that he or anyone else could ever receive an answer. Man, the primary agent of the change involved, is left out of account.[19]

Only in the most schematic way, then, should epistemes or paradigms be viewed as discontinuous. Such abstractions aid, and are in fact needed for, accurate analysis. But genuine clarity suffers if they are mistaken for the historical actuality. Nor do epistemes follow one another in neat sequence through time. Men may dwell side by side and yet work within different ones. Men may move back and forth from one to another before making a final commitment; some may never make one. And others, like Darwin, may promote one episteme while never breaking free of another.

Insofar as creationism and positivism shared certain scientific beliefs a common discourse was possible, and it was along the bridge of this discourse that men passed from one episteme to the other. What did they have in common? They shared the legacy of a joint scientific heritage: laws of nature (however differently rationalized), the imperatives of evidence, the canons of proof and prediction. In its pure form creationism predicted that no purely physical explanation of speciation would be found, that no transitional fossils would be discovered, that no argument for evolution could be constructed that would plausibly link supporting evidence. As will be seen, there were creationists who were more committed to a protopositivism within natural history than they were to the theological elements in their episteme, and these were the ones who, for the sake of their science, crossed the bridge; the others did not. Creationism, then, was not only a definite stance on the origin and nature of species, but was itself unstable owing to the development of positivism within it. And this was owing to the gradual modification of the tools of science through work.

When I speak of epistemes I mean those systems of ideas that are held by a group of naturalists working contemporaneously and which may be classified on principles of affinity. These systems, positivism and creationism, were realized in the lives of men, and that may best be understood by focusing on the reality of men thinking and working through time, out of one mode of discourse and into another. So much, I think, is true for the generation that makes the transition. Those raised within the confines of the new mode will be, of course, inclined to take its characteristics as self-evident, but only so long as it suits their needs. When conscious theory no longer fits their "silent practice" they, too, will inaugurate an epistemological shift.

The advent of evolution in natural history was the consequence of a change in the way in which science was thought about and practiced: The old scientific episteme, creationism, which mixed the Newtonian nomothetic and the Baconian inductionist traditions from the physical sciences with biblical theology and a type of philosophical idealism, had sanctioned, in the idea of special creation, or so it appeared from the new positive perspective, a pseudoparadigm that was not a research governing theory (since its power to explain was only verbal) but an antitheory, a void that had the function of knowledge but, as naturalists increasingly came to feel, conveyed none. This discontent with special creation was the result of a subtle and gradual shift in the epistemic foundations of natural history toward positivism.

The word positivism has been used in so many ways that it has ceased to have a clear and readily perceived meaning. Some have confined it to Auguste Comte's philosophy of science and the quasi-religious movement that grew out of it. Others have used it as a synonym for laissez-faire liberalism or as an all-encompassing term for naturalism and secularism. Some use it to represent one of a number of contemporary schools in the philosophy of science. This last is perhaps the most popular current usage.[20] In the face of such diversity, it would be folly to attempt a universally acceptable definition, even if that were possible. A writer's only recourse, I think, is to try to lay out clearly what he means by the term and hope that his readers will accept that usage provisionally as an aid in his exposition.

As used here, positivism signifies that attitude toward nature that became common among men of science and those whose intellectual lives were influenced by science in the nineteenth century, and which saw the purpose of science to be the discovery of laws which reflected the operation of purely natural or "secondary" causes.[21] It typically used mechanistic or materialistic models of causality, rejected supernatural, teleological, or other factors which were in principle beyond scientific examination as legitimate aspects of scientific inquiry, and—whatever the desires or beliefs of individual practitioners, many of whom were theists or even good Christians—embraced and promoted those far-reaching cultural tendencies conventionally known as secularism and naturalism. This development resulted in a consensus, virtually unquestioned among followers of science by the century's end, that science had a unique and prescriptive approach to knowledge which was superior to all others and was exclusively the way to understand the world of nature. The requirement that all questions touching nature be open to science, that there be no areas of knowledge about the world that were "forbidden ground," however, did not necessitate the assumption that all problems were practically open to solution. What the mystery of God was to the creationist, man's finitude was to the positivist. But while the former gloried in his inabilities, the latter had only a reluctant acquiescence.

The question of the development of positivism in science, and of science itself as a form of rationality, has recently come under such far-reaching inquiry that any attempt to account for it would be premature, if not pretentious. Rather, I shall take the growth of positivism throughout science, during the period under discussion, as a historical given. Its existence will not be questioned by many, I should think, and that agreement is all the present work requires.[22] The change of outlook in question had been in process at least since the seventeenth century. Formal thinkers played their part, of course, either as innovators or, more frequently, as catalysts, and I have no wish to denigrate their role, but the great ground swell of intellectual transformation was, I believe, the product of workaday scientific thought, that realm of activity ruled by what Abraham Kaplan has called "logic-in-use." The origins of this change in the way of viewing the world are obscure, complex, and fascinating. But whether discussed in terms of Foucault's epistemes or Kuhnian paradigms, whether traced to great philosophers or to social changes, one should not neglect the effect of the working man of science thinking through the problems of his craft on a day-to-day basis, testing his experience against received theory and improving or discarding that theory.[23]

Why did a shift to the positive episteme require a new theory of species origin? The problem that created a situation bearing at least a family resemblance to a "paradigm crisis" in natural history, though on a more elemental level, was not an "anomaly of discovery" or a failure of expectation in Kuhn's sense, but an epistemological anomaly, an anomaly in the understanding of science. Special creation, as an explanation of speciation, simply led nowhere when viewed from the perspective of positivism. In truth, it led nowhere within the creationist episteme, but there ignorance was acceptable as an end of the scientific quest. In the positive episteme it was intolerable. The need for a new species theory, then, grew out of the fact that naturalists were beginning to ask questions, formulated under positive assumptions, that could not be answered, or even asked, by creationism. It is often thought that positivism restricted the area of scientific inquiry by excluding questions of a metaphysical or theological nature. In this case, however, the struggle was over extending, by invoking natural causes, the number of questions that could be asked.[24] For creationists, a knowledge of the origin of species was not within the reach of normal scientific practice. But it was not so much that creationists could not give any answer to the question of how new species originate as that their answer was not meaningful within the positive understanding of what constituted a scientific solution: uniformity of law and natural causes. On this, as on other points, the positivists and the creationists were separated, in Kuhn's terms, by "incommensurable ways of seeing the world and by practicing science in it."[25] As naturalists came more and more to insist on positivistic standards and practice in science, the realization grew that some mode of evolution, of speciation by natural descent, was preferable to "sudden interruptions of natural order."[26] This realization was not always welcomed, but it could not be indefinitely resisted. In 1860, Thomas Henry Huxley could still treat both special creation and transmutation as scientific hypotheses. Yet he obviously felt the tension between positive science and any possible conception of special creation. Echoing Darwin, he asked, "How much, if anything does it explain; what does it really tell us about nature? Can such a cause even be discussed within the framework of science?"[27] That was the crux of the entire matter.

Its break with creationism should not lead one to identify the positive spirit in science with a faddish materialism or the optimistic cult of science and progress that flourished in the 1860s and 1870s. The episteme change was as compatible with human despair as it was with confidence. It could share the dark nights of the soul that afflicted many in the *fin de siècle* just as it fed the hopes of optimistic reformers. Positivism was not a social prescription but a way of practicing science and a conception of science as a way, *the* way, of looking at the world and man. The extent to which this outlook has lost (if, indeed, it ever had) self-evidence is a mark of the failure of science to achieve a general cultural hegemony. But in the area of nature, it was and generally still is unrivaled by any serious challenger.

To say that nineteenth-century science, including natural history, became predominantly positivistic in its epistemology is not, of course, to ignore the numerous philosophical controversies over the adequacy of positivism as a vehicle of knowledge or to suggest any subsequent diminution in their importance. Nor is it to pretend that scientists have ever been of one mind on the theoretical aspects of their pursuits or the philosophical implications of those pursuits. Some scientists today, as in Darwin's time, are atheists; some are theists; some are indifferent to such questions. Some follow Plato, some Kant, some Hegel, some Comte, some just ordinary common sense in their views on the problem of knowledge. Some play with abstractions never thinking that they duplicate reality; others take their models with serious literalness. But one and all, I think, follow in their day-to-day work the basic positive prescriptions of scientific practice. Theory grounded in experiment, in evidence and proof, is the hallmark of sound science, and this formula is based on assumptions of natural causes and the universal regularity of phenomena. It is this unquestioned, almost, in Foucault's sense, "unconscious," almost reflexive, theory and practice that I mean by positivism. While some might have thought that Charles Lyell overstated the case when he wrote in the 1872 and last edition of the *Principles of Geology* that "in truth there are only as yet two rival hypotheses, between which we have our choice in regard to the origin of species—namely, first, that of special creation—and, secondly, that of creation by variation and natural selection," he was certainly correct to narrow the choice to special creation and evolution.[28] The old theologically grounded science had led to the former, and the new positive science now led to the latter.

As models of positive science, physics and astronomy—and particularly the latter—were always before the eyes of naturalists in the early nineteenth century.[29] But it was geology, I suspect, that was the practical exemplar of the new orientation. Despite an emerging sense of divergence owing to specialization, geology was part of the kit of most serious naturalists during those years; and, while it is true that the leading geologists readily invoked a theistic philosophy to undergird their knowledge, as befitted those working within the creationist episteme, their practice was uncompromisingly positive. The manuals, handbooks, and reports of geologists were characterized by an emphasis on natural causes and empirical validation, and generally dispensed with mention of the Deity and his works. While this was not true of theoretical works—contrast, for example, Lyell's sometimes Paleyesque *Principles of Geology* with his dry *Manual of Geology*—or of works intended to present geology to the public, the presence of this positivism in books and articles intended for practitioners illustrates the developing tension within the creationist episteme that eventually would give rise to a positive reorientation in natural history. A sort of creeping positivism grew up in day-to-day practice within the old science and gradually proved its practical superiority. The famous war against Mosaic geology, fought by uniformitarian and catastrophist alike, is a familiar aspect of this development. But these allies in quest of an autonomous geology were not equally favored to flourish under a positive episteme. While catastrophism was perfectly capable of positive practice, its theory was marred by appeals to causes of kind and degree that, even if natural, were unknown in the present order of things and were, hence, inexplicable and therefore offered links to creationism. Uniformitarianism, on the other hand, by rejecting unknown and mysterious causes for explanations built solely on natural processes now in operation, was much more congenial to the new outlook. The shift in episteme, then, favored Lyell's geology just as it favored Darwin's biology. The often remarked points of similarity between the two were not owing merely to Darwin's discipleship. They both reflected profound changes in the understanding of the nature of science. Not the least of these changes was the denial to theology of a role in scientific explanation. This denial complemented other, far-reaching changes in western society.

That the development of positive science had a place in the growth of secularization in the nineteenth century has always been evident. What has not been so clear is just what that place was. It would, of course, be wrong to attempt to explain secularism as the result of the impact of science alone. Historians are now beginning to realize how complex an event secularization was. Some of the major determinants were not even closely related to science.[30] There is no way to approach the question of secularization, however, without taking into account the well-worn issue of the warfare between science and religion. The rise of the new episteme in science was basic to this problem and to any understanding of the broader cultural reaches of the conflict. The important thing to bear in mind is that secularism should not be thought of as solely an antireligious movement. It was a convergence of many processes and, insofar as these moved in opposition to traditional religion, conflict necessarily resulted. But these factors, positivism among them, at the same time moved toward a new reality, a new sense of the world and of man, and within that new reality a new parallel religious vision formed. To see secularization as a collapse of faith, therefore, is to put oneself in the way of never understanding it. It was not simply the old world emptied of faith, but the creation of a new world with new values. The conflict of science and religion was not a warfare so much as a transformation from which altered concepts of both science and religion emerged. Because the new episteme for science differed from the old in having within it no place for theology, and because an episteme is a communal definition of what is knowable, serious questions were thereby raised that made the conflict, sometimes dismissed as an illusion or a mistake, very real indeed. The important thing was to get theology out of science. The conflict, therefore, was both necessary and unavoidable if a positive science were to develop. The process of accommodation by which theology surrendered the world to science was often sufficiently voluntary and peaceful as to be easily confused with the absence of conflict. It is even easier to see it so in retrospect. But significant, and perhaps permanent, changes in mainline Christianity—both Protestant and Catholic—resulted from this readjustment of provinces of authority. Theologians have since pronounced their retreat a victory, and perhaps it was, but many involved at the time would not have thought so.

In some ways the conflict between science and religion that emerged in the nineteenth century was as old as antiquity. From the time of the earliest church fathers Christianity had existed in tension with contemporary secular learning. Everyone knows what had earlier happened to Socrates, to Aristotle, and, later, to the schools of philosophy under Theodosius. But these resemblances are superficial. The struggle that enveloped so many of Darwin's contemporaries, willingly or unwillingly, was radically different from those earlier battles over conflicting doctrines and theories.

With the full emergence of the positive episteme during the nineteenth century, science as a whole for the first time openly developed a completely natural world system, one that was neither logically nor theoretically obligated to theology in any way. The striking thing about science in the later nineteenth century—and which separates it from a similar positive movement among the mechanists of the seventeenth century—was the willingness of so many scientists, even pious ones, to dispense with the "God hypothesis" as a part of the presuppositions of scientific work. The separation of theology from science that developed in the nineteenth century had, of course, been potentially present since the seventeenth century. But, with some notable exceptions, almost everyone, particularly those in the tradition of Boyle and Newton, had been determined to prevent it.[31] Now, on the contrary, the fabric of unity was willingly rent. Why? It obviously will not do merely to invoke the rise of positivism as an explanation. That is only to explain a thing by itself. It is important to bear in mind what brought about the change in perspective among scientists. Owing to increasingly refined practice and to the increasing volume of successful science, practitioners began to feel that a theological domain in nature which neither yielded to nor permitted scientific investigation was an actual obstacle to the full development of scientific knowledge. Historians have increasingly and properly stressed that Victorian scientists, as a group, were not hostile to religion; that Thomas Henry Huxley and John William Draper were not typical in their polemics. This is true. But what must not be overlooked is that many Victorian scientists were uneasy or skeptical about the role of religion *within science,* and as the century wore on their numbers grew. This was the focal point of the conflict, and it turned on the question of knowledge. The episteme shift under consideration did not require the repudiation of religion as such. It only required its rejection as a means of knowing the world.

The old science, operating within the creationist episteme, had less difficulty about religion. The two admittedly distinct but not really separate modes of knowing, the scientific and the theological, were united in a common—and, by mid century, increasingly articulate—idealism.[32] While it is true that many creationists followed the accommodating Baconian tradition of insisting on a pragmatic ad hoc divorce of science and religion as a way of warding off conflict, they were, nonetheless, incapable of receiving a scientific teaching that appeared to conflict with theological truth. Their theology, willy-nilly, acted as a constant censor of scientific theory. A common truth was believed to run through the entire creation, and the perception of this truth was predicated on a strong theological belief. When the Duke of Argyll gave his inaugural address as chancellor of the University of St. Andrews in 1852, he warned his hearers of a new and dangerous error growing in popularity. "An absolute separation has been declared between science and religion; and the theologian and the philosopher have entered into a sort of tacit agreement that each is to be left free and unimpeded by the other within his own walk and province." Weak though the pact had been in the honoring (for scientists showed a regrettable tendency to violate it), it was based on a mistake. "The truth is," he said, "that no such separation between science and religion, as that which is often attempted, is possible. Every outward visible thing has some inward and spiritual meaning; and if the true one be not seen, a false one will be invented." What is now needed, he went on, is neither a "compounding of physical with religious truth nor . . . a total separation between them," but "a true idea of the nature of their connection. . . ." T. Vernon Wollaston, in reviewing the *Origin of Species,* acknowledged the wisdom of avoiding a contact between the two realms, but concluded that "no man who loves truth, in all its phases, for its own sake, will long rest contented in accepting *as such* a zoological creed which is in direct antagonism with his theological one." One must fall, and while Wollaston refrained from deciding which one in this instance, no reader could have mistaken his favorite. Science, then, might go its own way up to a point, but it ultimately had to return to the one truth: the presence of Mind in the world, a presence demonstrable by science and theology alike. Defending the popular idealist doctrine of purposeful design in nature by appealing to an analogy with human purpose, paleontologist Richard Owen argued that "the necessity . . . of our intellectual nature compels us to test the cause of type and correlation [in nature] by the human standard." He roundly condemned the new episteme as not only wrong, but perverse. "The result of such a test," he continued, "is only not knowledge [*sic*] through a special and restrictive definition forbidding its contribution, as such, to the general stock of mental assets. But the human mind never has submitted, and probably never will submit, to so positive a ban." He likened the new "incapacity" to perceive mental analogies in nature as "possibly more abnormal or exceptional" than the ability to detect them. "I have, indeed," he concluded, "been led to liken it . . . to a case of colour-blindness."[33] It was indeed—a mutual blindness that is clearly seen in the differing conceptions of laws of nature.

The gradual movement from an idealist conception of a law of nature to a positive one, from law as divine will to law as no more than observed regularity of behavior, was a major point in the episteme shift, and one that was reflected in Darwin's continuing frustration over the relation of the Creator to the world. For Adam Sedgwick, one of the foremost British geologists of the century, the laws of nature were meaningful only if they were thought of as acts of God. While realizing that such laws were, in practice, the formulations of men, Sedgwick held that if one denied or ignored the role of God in the world order one could not logically formulate any rational system of natural law because the necessary theological and metaphysical basis of such a formulation was absent. "Law," he wrote, "implies a lawgiver, and without that notion the word law is without meaning." Within Sedgwick's episteme this was quite true—a tautology, in fact. To the positivist, on the other hand, such thinking as this was merely a senseless play on words. He had an entirely different idea of natural law which rested on a totally different conception of science and of human knowledge.

As a creationist, Sedgwick was deeply offended and threatened by Robert Chambers's notorious evolutionary treatise *Vestiges of the Natural History of Creation* (1844). Not content with a three-hundred-page demolition in his preface to the fifth edition of his *Discourse,* he returned to the attack in a supplement to the appendix at the end of the volume. However bad Chambers's science may have been, no one can read these pages and conclude that Sedgwick's indictment of pantheism and materialism, his outrage at the idea that speciation by natural generation was the only proper theory open to the naturalist, was merely a matter of scientific accuracy. Sedgwick's science and his theology were bound up in a unified system such that an attack on one was necessarily an attack on the other. The same was true of Louis Agassiz, Richard Owen, and other idealists. When Sedgwick accused Darwin, on the publication of the *Origin of Species,* of trying to break the chains that bound final causes to secondary ones he was right; and when he described this act to another as embodying the intention of making "us independent of a Creator" he was seeing it in the only way that he could. In the words of the *Princeton Review,* the positive enterprise was one that "avowedly and purposely ungods the universe."[34]

Creationists like Sedgwick, then, found it impossible to think of nature without thinking of God. To theists who worked within the new episteme the relation of the divine to nature was not so self-evident. British physiologist William B. Carpenter pointed up the tension between theology and positive science when he wrote defensively that "we do not exclude the subsequent perpetual agency of Creative Will, because in scientific reasoning we speak of it in the language of physical force." Admitting that "the idea of 'government' by a God" had been properly excluded from science, he protested against substituting an autonomous government of nature by laws. Laws, he argued, must have a cause and, while science could not reveal that cause, it could not exclude the idea of "an Intelligent First Cause" as their source. The uncomfortable fact was, however, that the new science had no use for such a *causa causarum.* For positivists like Carpenter it was possible to retain God as part of one's personal view of the cosmos, but such retention was private, subjective, and artificial. One could keep God in the positivist cosmos only by constantly reminding oneself that he was there. Unlike the creationist cosmology wherein he played a specific role as creator, innovator, preserver, and designer, in the positive episteme there was no equivalent function for him to carry out. He was, at best, a gratuitous philosophical concept derived from a personal need and not entailed by the new system of science. Neither its rationale nor its logic required his presence to get on with scientific work. Consequently, many theists, such as Carpenter and American botanist Asa Gray, were never entirely easy within the fold of the new science.[35]

The dilemma of the theist who practiced science within the new episteme was not the least of the various and complex changes that promoted a religion of faith which paralleled the acceptance of a positive science. Just as science shifted from a theological ground to a positive one, so religion—at least among many scientists and laymen influenced by science—shifted from religion as knowledge to religion as faith. The attempts to defend science from the charge of irreligion, which were so common among this group, necessarily required real, if subtle, redefinitions of religion and especially of Christianity. This, of course, involved no novel approach to either theology or the Scriptures. Prescient theologians had prepared the way well beforehand. The Bible as history had gone the way of the Bible as science and had given way to the Bible as myth. Myth was thought of, not as error, but as symbolic theological truth. Ethics replaced ritual purity and behavioral conformity as the center of concern. Christianity became hard to distinguish from theism and theism from a vague belief in some sort of spiritual dimension in life. The influence of scientists in weaning a large portion of the public away from biblical literalism and changing the nature of that public's understanding of Christianity should not be underrated and certainly needs more thorough study.

But was this accommodation by theistic scientists necessary? It might be granted, given suitable definitions of terms, that, as Owen Chadwick has said, "the advances of science could hardly touch God," that "no enquiry in the realm of the physical could produce results in the realm of the spiritual."[36] This, however, presupposes that a separation of religion and science has already occurred, but it is a separation that was nonsensical to the creationist. The issue raised by positivism, as a theory of knowledge, was whether there *was* a realm of the spiritual. Certainly the new episteme had no place for such an idea. Logically, that is, it had no place for one. But few practitioners, including Charles Darwin, were willing to go so far. Hence, while religion was banished from the province of their science, which was the physical world—past, present and future—it was not banished from their lives and, taking a form there seemingly not relevant to science, often flowed back in curious ways to influence their science. American biologist and life-long theist Edward Drinker Cope looked on it as one of the true functions of science to separate the realm of knowledge which was proper to the understanding from that which was peculiar to the spirit; and the latter, he expected, would be "reduced in a large part to *personal religion.*" Science, he confidently wrote to his wife, will, in time, "straighten up the theologies." At the same time, Cope was not atypical in never losing his faith that, somehow, God lived and ruled the creation.[37] But scientists generally abandoned the attempt to give scientific proof from nature of God's existence. Apologetics based on design in nature, an enterprise that had not only made sense under the old episteme but was logically mandated by it, now to many no longer seemed either necessary, proper, or even possible.

One might go so far as to suggest among these persons the existence of a third episteme: one that divided truth into two portions and endowed man with dual faculties of perception, that continued to ask the old questions about creation, cosmic purpose, and God, but attempted to give answers compatible with the science of the new episteme. Both the creationist and the positivist, in their pure forms, would have rejected such a compromise, and attempts of this kind did fail to become a lasting part of the new biology, but many troubled minds found comfort in some such alternative. Certainly the postulation of the existence of such an episteme would be preferable to the continued evasion of the fact that so many scientists during this period were religious by attributing their piety to public mollification and polite duplicity, on the one hand, and to the mistaken belief that there was no problem between science and religion on the other.[38]

When religion was a noisier question, however, it was not religion, as such, that was at issue but rather the not always welcome theological associate of the creationist episteme: that mixture of ecclesiasticism, biblical literalism, and intellectual and social conservatism that went by the name of orthodoxy among its foes and, among its friends, was known (without complete inappropriateness) as Christianity. Agnostics and theists alike among the positivists were unwilling to acknowledge the legitimacy of the latter title. Huxley, for example, denounced orthodoxy as an "idolatrous accretion" on the body of genuine Christianity, and Carpenter thought that reflective believers were abandoning an increasingly incredible theology. Sebastian Evans, moving to the attack in the pages of *Nature,* spoke in tones of extreme bitterness of "the felicitous instinct of clerical antagonism" which had led an orthodox critic of prehistorian John Lubbock to make his attack where Lubbock was strongest and to lay "about him with all the fine, fervid imbicility distinctive of his particular clique." When the intellectual orthodox gathered themselves into the Victoria Institute to promote, in effect, the creationist episteme, contempt and derision were the responses of positivists. The discussion of the *Origin* at one meeting, Darwin wrote to Alfred Russel Wallace, was "very rich from the nonsense talked." Geologist T. G. Bonney found the misrepresentations of scientific work in this body "so gross and so palpable, that assuming them to have been made in good faith, . . . indicate either an extraordinary amount of stupidity or a singular form of mental perversity."[39] These harsh words were not untypical, nor were they unreciprocated. The vituperation is best understood when it is seen, not simply as a clash between science and religion, but as one between two antagonistic scientific epistemes: one with a deep theological commitment, the other with none. Scientific popularizer John Fiske was impressively correct when he scolded polemicist John William Draper, saying that the real conflict was not between science and religion, but between two systems of science.[40] Neither the positivists nor the advocates of the elusive third episteme could allow theology to lay claim to a knowledge of nature; and no more could orthodoxy, in close alliance with creationism, admit to an ignorance of it.

Such, then, roughly blocked out, is the conceptual framework within which I will try to analyze Charles Darwin's mental world as it related to science and, specifically, to natural history. The categories used herein are exploratory and, like all ordering principles, may illuminate to a point, but end in obscuring some things, perhaps many things.

2 Special Creation among British and American Naturalists, 1830–59

Charles Darwin's hostile preoccupation with the belief that God had separately and individually created each of the animal and plant species in the world is one of the most intriguing but neglected features of the *Origin of Species*. Historians have disagreed about what to make of it. Some have dismissed it as an eccentricity born of Darwin's isolation at Down and his consequent ignorance of the fact that his fellow naturalists had all but abandoned the notion. Others have implied that Darwin kept up with things, but was simply naive: he thought that when scientists spoke of creation they meant it literally in the miraculous sense of the Old Testament. Some have accused Darwin of setting up a straw man in order to improve the appearance of his own case. Lastly, there are those who believe, correctly I think, that Darwin's rejection of special creation was part of the transformation of biology into a positive science, one committed to thoroughly naturalistic explanations based on material causes and the uniformity of the laws of nature, a change to which the *Origin* was a signally important contribution.

I hope to show that the first three interpretations rest on a failure to appreciate the subtlety and extent of the opposition that Darwin faced among his fellow naturalists—not only from convinced special creationists, but from those who, while they may have developed doubts about special creation, still worked under its inhibitions and discouragements. That natural history, as it was related to the concept of species and the history of life on the earth, had reached a crisis in the 1850s was not an uncommon belief among naturalists at that time. To a great extent, the crisis signified the epistemic breakdown of creationism as a vehicle for scientific work. The legacy of this disintegrating creationism was an attitude of nescience on the species question and of indifference and hostility to theory, outlooks that were blockages to the development of a positive biology as serious as creationism itself and which were related, in ways not always clear to contemporaries, to the assumptions of creationism. Similarly, the "ordinary argument" for special creation with its emphasis on benevolent and rational design in nature, which Darwin constantly attacked in the *Origin,* was not, as W. F. Cannon correctly notes, in a strict modern sense, a rival scientific theory.[1] It rather represented a persuasion, an atmosphere that permeated natural history so universally that naturalists were often unaware of the extent of its influence. Consequently, it was not a harmless straw man, but a traditional bias found among scientists and laymen alike and one that stood in the path of any novel way of viewing the problem of species. Darwin, then, was not engaged in anachronistic shadowboxing, but had selected his target well and knew exactly what he was doing. His attack on special creation was a response to the crisis and an attempt to resolve it by helping to promote the restructuring of biology along positivist lines. The critique of special creation in the *Origin* was systematically organized to that end. I also hope to show the extent to which his critique was a theological response, and the degree to which theological concerns affected his science without being directly a part of it.[2]

The concentration on naturalists in Great Britain and America is admittedly restrictive and, to some extent, artificial, but it has two justifications. First, the intellectual communities of these two countries made up the arena in which Darwin personally fought. They constituted the milieu in which he matured and which he, personally and most directly, sought to transform. Second, Darwin's attack on special creation was closely related to the allied question of biblicism in science. By "biblicism" I do not mean biblical literalism or "fundamentalism", but the continuing employment of biblical, especially Old Testament, images and language in science, and of explanations directly or indirectly grounded in Christian theology. While this was not a unique element in Anglo-American science by any means, nor as strong as it once had been, it was still a prominent feature in some quarters at mid century. The continued use of these traditional motives, even by those who had given up special creation, and even among Darwin's converts and followers, is an important although separate inquiry from that undertaken here. It certainly deserves study. For present purposes, however, it must be sufficient to note from time to time the connection between biblicism in scientific writing and thinking and the creationist outlook and its residue. The positivist aversion to biblicism was one of those conventions of discourse that characterizes and helps to define an intellectual community, that, in Foucault's words, governs "its procedures of exclusion."[3] Further, in treating Anglo-American natural history as a virtually undifferentiated whole, I do not intend to suggest that there were not important areas of divergence. On the contrary, I will note two: the apparently greater belief in miraculous creation in America, and a corresponding lesser enthusiasm there for the aggressive ideology of positivism. I think, however, that common themes may be stressed for the purposes of this book.

By special creation I shall mean the belief that God in some way *directly* intervened in the order of nature to originate each new species. This view was not synonymous with the position ridiculed by Darwin as the idea of "elemental atoms [being] commanded suddenly to flash into living tissues" by supernatural agency,[4] but as it was held by some prominent naturalists, it did come surprisingly close to that. More frequently, those holding creationist ideas would plead ignorance of the means and affirm only the fact. The creative act could be either miraculous or the result of some unknown lawful process involving unknown causes (natural or preternatural) privy only to the Creator. The important factor was God's personal and purposeful involvement. Hence the doctrine of special creation was associated with a second idea, the doctrine of design, or the belief that there was in the workings of nature compelling evidence of the operation of intelligence moving to achieve certain predetermined ends. A third factor in this cluster of ideas was the belief in the stability of species. Each species was a unique, discrete, and absolute creation. It might vary slightly under altered environmental conditions, but its essence never changed. It appeared, flourished, and became extinct in accord with the will of the Creator.

Schematically viewed, special creation took two forms in the scientific community.[5] One is easily identified by its proximity to the biblical tradition in language and imagery. The other was within the Newtonian convention of natural law in which the Creator was thought to employ certain laws of nature to work his will. This was not supernatural creation in the miraculous sense, but one nevertheless in which God was thought to be directly participating, guiding and supervising each natural event. Divine intervention, in this sense, was a purposeful act, compatible with the ordinary laws of nature, and done to accomplish a definite end. Unlike the operation of such continuous laws as gravitation, however, creative law was discontinuous in its application and responded to a periodic divine initiative. To a modern, the distinction between lawful and miraculous creation may seem more verbal than real, but it was an important distinction to those who made it. The rationality of their science, and its ability to construct an ordered and lawful history of life, depended on it. Accordingly, these creationists usually denounced the miraculous creation of new species with appropriate scientific fervor. Their talk of law, however, should not mislead one into thinking that their explanation of the origin of new species was naturalistic, that is, positive or nontheological. This variety of the concept of creation by law, or nomothetic creation, as I shall call it, appealed to naturalists who leaned toward a law-bound biological science. Charles Lyell, for instance, in the *Principles of Geology* (1830) spoke in the Newtonian manner of "general laws which may regulate the first introduction of species . . . [and] may limit their duration on the earth."[6]

Such a view was acceptable to the special creationist, however, only if law were an immediate instrument of the Creator. Some, accordingly, viewed the generic idea of creation by law with distrust, and not without cause. Too far removed from God's directing hand, as in some transmutationist doctrines, such a system of laws might function automatically and exclude God from any direct activity in the world. American geologist Edward Hitchcock, as late as 1859, saw all attempts to establish the idea of creation by law, that is, the origination of species by inherent forces or tendencies within nature, as attempts to expound atheism, even when advocated by theists or Christians. Such beliefs, he claimed, "virtually annihilate the doctrines of miraculous and special Providence and of prayer." British geologist Adam Sedgwick shared Hitchcock's suspicion that attempts to account for new species by means of "secondary laws" were inherently materialistic and atheistic: "There is no material law," he wrote in 1850, "to explain the beginning of life, sensation, and volition." In 1860 Sedgwick told paleontologist Richard Owen that "it is clear that there has been a law governing the succession of forms. But here, by law, I mean only of succession, and not a law like that of gravitation, out of which the actual movements of our system follow by mechanical succession. In that sense, I do not believe in any law of creation." Such forms could derive only from an ideal archetype and were manifestly the products "of the action of a final cause—i.e. of intelligent causation, or creation." Lyell's friend and colleague, Gideon Mantell, had earlier asserted that while the physical world was governed by law, geology "does not affect to disclose the first creation of animated nature; it does not venture to assume that we have physical evidence of a beginning: *it does not warrant the attempt to explain the miraculous interventions of Providence, by the operation of natural laws.*"[7]

The core of special creation, then, was not miracles (although individual naturalists did appeal to them), but the direct, volitional, and purposeful intervention of God in the course of nature, by whatever means, to create something new.

The ambiguity of the creation by law concept, which, in practice, could and did mean anything from the belief that God works through appointed laws to make new species on the one hand to transmutationism on the other, introduces a serious problem that must be faced in any study of special creation. That is the use of the term creation by naturalists in the mid-nineteenth century.

In general, the term had four distinct meanings. First, it was meant literally, i.e., that God miraculously made a species. Hugh Miller, the popular Scottish geologist, explaining the first appearance of fish in the fossil record, wrote:

> Nature lay dead in a waste theatre of rock, vapour, and sea, in which the insensate laws, chemical, mechanical, and electric, carried on their blind, unintelligent processes: the *creative fiat* went forth; and, amid waters that straightway teemed with life in its lower forms, vegetable and animal, the dynasty of the fish was introduced.

Each new epoch in life was introduced *"through an act of creation."* Edward Hitchcock, in the 1858 revision of his tremendously popular *Religion of Geology,* argued that nature was the result of God's creative power "to the exclusion of every other cause." The fossil record of life showed not a continuous development but a discontinuous start-and-stop process indicating repeated divine interventions. Admittedly, God used instrumentalities to create: we are told in Scripture that the earth brought forth grass, and so forth. But his power was behind this—and not far behind! God, Hitchcock went on, could and did work miraculous interventions in nature: "Geology shows a divine hand cutting the chain asunder at intervals, and commencing new series of operations." The Creator's "miraculous fiat" was the source of life. Against the transmutationists Hitchcock argued that "for the most part" new races came in as groups as old ones went out. This pointed "clearly to creation rather than development." "No other science," the geologist boasted, "presents us with such repeated examples of special miraculous intervention in nature." Louis Agassiz, who dominated American natural history in 1859, wrote in his *Essay on Classification* (1857) that of the three possible modes of the origin of new beings—spontaneous generation by the operation of natural laws, the operation of creative laws established by God, and "the immediate intervention of an intelligent Creator"—only the last was supported by science, albeit the means by which God created was unknown. Agassiz's student, geologist Joseph LeConte, in the same year, asserted that "as far as the evidence of geology extends, *each species was introduced by the direct miraculous interference of a personal intelligence.*" Agassiz and Augustus Gould, in their 1848 zoology textbook, invoked a simple biblical creationism in which species were made by "the direct intervention of creative power."[8] Examples like these are difficult to interpret as anything other than assertions of sudden, unique, mysterious, and literally miraculous acts of creation involving means utterly beyond the laws governing physical nature and beyond human understanding. References by others to an act of creation are less clear. Some are obviously not to be so interpreted.

When Charles Lyell wrote, regarding changes in animal populations, that there were cases in which "the animals inhabiting any given district have been partially altered by the extinction of some species, and the introduction of others, whether by new creations or by immigration," and, elsewhere, that the peculiar flora and fauna of St. Helena seemed "to have been expressly created for this remote and insulated spot," it is by no means clear that he intended to invoke a miracle: given what we know of his philosophy of science, of his insistence on uniform natural law and his dislike of biblicism, we may be sure that he did not. Lyell was employing creation in the second conventional sense, which was that favored by nomothetic creationists: as symbolic of an unknown mode of divine action that stayed within the confines of the laws of nature and, while involving a continuing divine initiative, did not require disruptions of that order. This creative process was no less mysterious than miracle, only less capricious. This was probably the meaning intended by the young Darwin when he entered in his *Beagle* diary (1834): "It seems not a very improbable conjecture that the want of animals [around Valparaiso, Chile] may be owing to none having been created since this country was raised from the sea." In the *Journal* (1839) he remarked that "some authors" (he was not among them by this time) explained the "American character" of the inhabitants of the Galapagos Islands as the result of "the creative power [having] acted according to the same law over a wide area." Lyell reported physicist John Tyndall as remarking in the 1850s that defining "creation" to mean some presently unknown "*modus operandi*" or "plan of introducing new species . . . might be a legitimate mode of dealing with the difficulty or mystery." But Richard Owen's famous 1858 definition of creation as "a process he [the zoologist] knows not what" received only ridicule from Darwin. When creationist language had reached the point where it was admitted even by its users to be empty of scientific content it served merely to manifest and not to resolve the mystery. Some openly embraced this function of the term. Creation, wrote mathematician-geologist William Hopkins, "is a negation of other theories rather than a theory of itself, and therefore cannot be called upon to account for phenomena at all in the physical sense. . . ." Its content was essentially negative: "a mode of expressing our disbelief in the assertion that such phenomena are, like ordinary phenomena, due solely to the action of ordinary natural causes, together with our belief that there is some higher order of causation in nature." An act of creation, Hopkins elaborated, was "a result due to some cause beyond and above those secondary causes to which the ordinary phenomena of nature are considered to be due." Ordinary phenomena were marked by "continuity" and the extraordinary by "discontinuity." Normal birth and growth was a continuous process. Transmutationists said that speciation was also a continuous process; but special creationists saw it as discontinuous, obeying as it did a higher law, unknown to science, which, while acting in conjunction with continuous nature, worked according "to some self-imposed law of the Divine Mind."[9]

As late as the 1850s, however, Darwin and others who had dispensed with special creation in either of its forms were still using creationist language. What meaning had it for them? It was the third sense: simply a conventional term meaning the origin of species, the world, matter, or whatever, and had no theological referent.[10] In his 1845 edition of the *Journal,* long after he had become convinced that species originated by descent with modification, Darwin referred for the first time to the Galapagos species as "aboriginal creations," but by this he meant only that they were peculiar to the island group. In the 1844 sketch of his theory of descent with modification he spoke of species being "created or produced," or "formed." As late as 1856, in writing to Lyell, he could refer to "our recent species" being "created." J. D. Hooker, who had all but embraced transmutation, must have been using the term in this neutral sense when, in commenting on the manuscript for chapter 11 of Darwin's *Natural Selection,* he wrote, "I am against making arctic regions centers of creation either by variation or by specific creation." When Darwin wrote of "single creations" in that same manuscript he meant only that each species is unique in time and place and not duplicated elsewhere. Reviewing Henry W. Bates's book on mimickry in the *Natural History Review* in 1863, he wrote, "It is hardly an exaggeration to say, that whilst reading and reflecting on the various facts in this Memoir, we feel to be as near witnesses, as we can ever hope to be, of the creation of a new species on this earth."[11]

A fourth meaning, which was confined to transmutationists and hence relatively rare until after 1859, was employed by Robert Chambers in his *Vestiges of the Natural History of Creation* (1844). Here creation referred to the body of laws established by God in the beginning, through the operation of which new species were evolved. For Chambers, each new species was a divine creation, but did not involve any special intervention or initiative on God's part as was the case for the nomothetic creationist.[12]

These examples, which could be multiplied in each of the four categories, show the problem faced by anyone attempting to estimate the strength of special creationism in natural history in 1859. The use of the word creation in itself means nothing. Its meaning can be determined only by the context in which it appears and by what we know independently about the beliefs of the author in question. Interpretation is further complicated by the fact that the same author may use the term in different ways during his career. Hence, we cannot assume that when, say, Lyell said "created" he always meant the same thing. Even in the mature Darwin there are rare instances where creation may be meant in the sense of a direct divine act. The following assessment, then, may contain errors of interpretation in individual instances. But, even allowing for them, I believe that the continuing obstruction presented by special creation to any scheme of development may be established.

How prevalent was a belief in special creation among naturalists in 1859? Despite Darwin's polemics in the *Origin,* it is well to remember that while a belief in the immutability of species did not logically require a belief in either form of special creation, immutability did obviously correlate strongly with antitransmutationism. The exclusion of transmutation left either special creation or nescience on the question of how species did originate virtually the only theoretical option open to naturalists. Of all of those mentioned by Darwin in 1859 as being foes of the mutability of species—Cuvier, Owen, Agassiz, Barrande, Falconer, Forbes, Lyell, Murchison, Sedgwick—apparently not one, with the exception of Agassiz, (Cuvier and Forbes were dead, of course) believed in miraculous creation in 1859.[13] All of these men, however, were paleontologists or geologists, and special creation, whether miraculous or nomothetic, was commonly recognized by them to have strong empirical evidence in the fossil series which seemed to support the idea that species appeared full-blown suddenly, endured unchanged, and became extinct without leaving descendents. Equally important, however, the fossil record gave no clue as to how species were created. To provide such a clue and to overcome the weight of negative evidence was a large part of Darwin's task.

The assumption that a belief in special creation, in whatever form, was needed to explain the fossil record was, then, the linchpin that held together the opposition to transmutation. Writing against Robert Chambers's *Vestiges of the Natural History of Creation* in 1846, the *American Journal of Science* declared, "Geology, if its facts mean anything, fully shows that tribes of animals have successively disappeared, owing to physical causes; and that the new races have appeared by creation, and not by graduation, or 'progress.'" James Dwight Dana, second only to Agassiz in influence among American naturalists, concurred. The geological record, he wrote in 1856, shows "successive creations of species, in their full perfection." It provided no basis for "presumptuous hypotheses" as to means. In the same year English entomologist T. Vernon Wollaston rejected transmutation as contrary to both known facts and sound reasoning. William Hopkins, addressing the Geological Society of London as president in 1851, had assumed as a matter of course that neither uniformitarians nor their opponents held "any notion of the transmutation of species." And by the time Huxley reviewed the *Origin* in the *Times* in 1859, he could treat special creation as a corollary of the immutability of species, so closely were the two conventionally associated. In attacking special creation, Darwin was not assaulting a moribund theology but a living and powerful idea.[14]

The major tendencies of Anglo-American naturalists in dealing with the problem of the origin of species in the period just prior to 1859 fall, schematically, into several groupings similar to and, of course, related to, those on the meaning of creation. Close to the still powerful biblical tradition in conceptualization and, usually, in personal religious faith, were those who felt confident that the evidence of science confirmed their belief in a miraculous creation. Advocates of this position, while certainly a minority and declining by 1859, were nonetheless leaders in their respective fields and by no means cranks. Miller (d. 1856), Hitchcock, and Agassiz were among the most prominent of these naturalists. The views of Miller and Hitchcock, already noted, need not be repeated. Agassiz, however, enjoyed exceptional notoriety in circles opposed to his belief as well as among its friends, and will repay further examination.[15]

In his Graham lectures in Boston in 1862, the Harvard scientist endorsed both an extreme catastrophic view of the earth's geological past and the doctrine of subsequent creations.

> The earth has been again and again inhabited by different successive generations of different kinds of animals, with interruptions between them, indicating great disturbances in the natural course of events and extensive changes in the prevailing conditions through which the earth has passed, accompanied by successive renewals of its inhabitants.

These creations might have been done, he acknowledged, either by an immediate act or by law. He clearly favored the former. Animal structure, he wrote,

> is not a kind of work which is delegated to secondary agencies; it is not like that which is delegated to a law working its way uniformly; but is that kind of work which the engineer retains when he superintends and controls his machine while it is working. It is evidence of a Creator constantly and thoughtfully working among the complicated structures that He has made.

In his *Essay on Classification* Agassiz marveled that so many who believed that God instituted the laws governing physical nature could not accept divine intervention in the living creation. The laws of physical nature, he argued, could not explain the existence of living things whose creation and duration on the earth were explicable only as the result of immediate divine action. He claimed no knowledge of the unique creative process, for analogies based on observable means of reproduction could give no guidance. The present inability of science to explain this mystery of nature was itself evidence of the reality of the divine miracles. But although unable to explain creation, Agassiz did think that science could witness to it. American geologists, he claimed, had actually discovered "the exact strata which contained the fossil remains of the first created animals" in the Lake Superior formation.[16]

Agassiz's creative excesses did much to further discredit among scientists the already failing idea of miraculous creation. His improbable taxonomic multiplications (400,000,000,000 new species of fish, William James satirically reported on the Thayer expedition to Brazil in 1865–66); his extravagant glacial theories and Hollywood-style mass creations were the reductio ad absurdum of miraculous special creation. No critic could have improved on it. Hooker described him to fellow botanist Asa Gray in 1854 as an "extraordinarily clever fellow and a treasure as a scientific man" but dismissed his theories as "too extreme for respect and hence . . . mere prejudices." Owen, though not referring directly to Agassiz, found miraculous creation "by the very multiplication of its manifestations" to be incredible. Lyell wrote to George Ticknor in 1860 that he had been driven "far over into Darwin's camp" by Agassiz's *Essay*.

> For when he attributed the original of every race of man to an independent starting point, or act of creation, and not satisfied with that, created whole "nations" at a time, every individual out of "earth, air, and water" as Hooker styles it, the miracles really become to me so much in the way of S. Antonius of Padua, or that Spanish saint whose name I forget, that I could not help thinking Lamarck must be right, for the rejection of his system led to such license in the cutting of knots. . . .[17]

And yet, five years later, Darwin acknowledged that there were still "not a few" who, like Agassiz, used the multiple creation of species to oppose his theories. Despite the tendency of Darwinians then and since to treat him lightly, Agassiz was not a clown and the opposition he represented was not a joke.[18]

A second tendency regarding specific origin was closely related to the nomothetic creationist's definition of special creation as the manipulation of natural laws by the creator. Naturalists of this persuasion who were convinced that speciation was a lawful process differed from Agassiz and his followers in avoiding the open declaration of a miraculous creation of species but treated the event as equally mysterious. The whole question of origins was, in effect, a problem that was beyond the present reach of science. This option is one of the most difficult to identify since it was more an attitude than a doctrine. It was a blend of scientific ignorance, piety, and discouragement that men of science seem to have drifted into and out of as their confidence waned or waxed. It might rest, like scientific nescience to be discussed later, on a Baconian reluctance to speculate without facts (a manifestation of the contemporary reverence for "induction"), or on the allied belief that the beginning of a species was unobservable and hence beyond inquiry. But it was always associated with theological beliefs. In fact, those who embraced this form of nescience on the species question coupled it with pious assurances of the operation of some sort of divine creative force or power, though denying the present possibility of sufficient scientific knowledge to even wonder as to the manner of action.

This discouraging appraisal was characteristic of Richard Owen, England's foremost comparative anatomist, in the 1850s, although he later shifted to a more confident stance. Writing in a popular scientific work, *Orr's Circle of the Sciences,* in 1854, he told his readers that no evolutionary theory could withstand examination and that we must, for the present, rest confident in the reality of "creative acts," though ignorant of the means. In 1859 he told a Cambridge University audience, "On the problem of the extinction of species I have little to say; and of the more mysterious subject of their coming into being, nothing profitable or to the purpose." He reviewed current speculations and dismissed them all. "Past experience of the chance aims of human fancy, unchecked and unguided by observed facts," he cautioned, "shews how widely they have ever glanced away from the gold centre of truth." The previous year, in his presidential address before the British Association for the Advancement of Science at Leeds, Owen had assured his audience that even if scientists should discover the laws governing the origin of life, "we should still retain as strongly the idea, which is the chief of the 'mode' or 'group of ideas' we call 'creation,' viz. that the process was ordained by and had originated from an all-wise and powerful First Cause of all things." In the meantime, the term "creation," which signified an unknown process, was both a confession of ignorance and a call to humility. "This analysis of the real meaning of the phrase 'distinct creation,'" he concluded, "has led me to suggest whether, in aiming to define the primary zoological provinces of the globe, we may not be trenching upon a province of knowledge beyond our present capacities; at least in the judgement of Lord Bacon, commenting upon man's efforts to pierce into the 'dead beginnings of things.'"

The word "creation," wrote Sedgwick, who exceeded most nomothetic creationists in his nescience as well as in his divergence toward miracle, refers to the primeval creative acts of God which were not bound by the present laws of cause and effect. It "expresses the exact content of our knowledge"; it signifies our ignorance and our belief in divine acts: "we can form not the least conception" of how God creates. The causes of species, he continued, even if natural and involving secondary causes, are not "*ordinary* natural causes; but . . . such as have not been ever submitted to the reason and senses of man. They belong not, therefore, to the province of inductive science." "Creation," concurred Oxford geologist John Phillips in 1860, "admits of no explanation in human language, because it refers to an act of God's power transcending all human thought and experience." Philosopher of science William Whewell, in 1857, also put the question of the origin of species beyond the reach of science: "Geology is silent," he wrote. "The mystery of creation is not within the range of her legitimate territory; she says nothing, but she points upwards." "The method of creation of a living species," wrote Dana in 1858, "appears now, more than ever before, to be a subject beyond the pale of human research."[19]

It was not a coincidence that catastrophist geologists such as Sedgwick, Dana, and Phillips were found in this camp. While it would be impossible, after recent scholarship, to deprecate the merit of their geological work as science and to treat it as merely a form of theology or biblical literalism, it should, nonetheless, be recognized that their methodological presuppositions precluded a full explanation of geological and, especially, of paleontological phenomena within the present order of nature. Indeed, it was in paleontology, in explaining the sequence of fossil forms of life, that their science most clearly revealed its theological associations. Their emphasis on the discontinuities in the geological past, on gigantic natural forces operating on scales not subject to present-day empirical study, was psychologically of a piece with the theoretical blank of creationist speciation. Both left the scientist to accept his ignorance as best he could. There was in some—one thinks of Sedgwick and Agassiz—a pious gladness in this inability to probe to the depths the secrets of nature, as if God's being was glorified in man's weakness.

This nescient approach to the problem of origins, while normally accompanied by a religious faith no less thoroughgoing in its devotion to the divine creative act than those who frankly attributed new species to miracles, was, yet, in a sense, a halfway point on the road to a fully positivistic science of biology. Its advocates accepted (with whatever reservations) our second definition of special creation: that it happened within the confines of natural law. But the awesome involvement of God in such unique and mysterious events inhibited scientific cognition. With the commitment to law it took one step toward Darwin, but only one. This conclusion, which invoked a lawful but mysterious and probably unknowable means of creation, was almost as inhospitable to Darwin's work as miraculous creation.

Perhaps the most well-known indication of a general shift of mind from miraculous creation to nomothetic creation is John Herschel's often-quoted 1836 letter to Lyell, news of which so delighted and encouraged young

Darwin. "The origination of fresh species," Herschel wrote, "could it ever come under our cognizance, would be found to be a natural in contradistinction to miraculous process—although we perceive no indications of any process actually in progress which is likely to issue in such a result." Lyell agreed that new species came in "through the intervention of intermediate causes" and complained of being accused of advocating miracles because he refrained from bluntly saying so in the *Principles* out of respect for religious sentiment. Even so, Lyell clearly required a stage-manager role for the Creator in this "natural" process, as, presumably, did Herschel, who also envisioned the "natural" process as an instrument of the Creator. The challenges presented by a piecemeal creation which required the fitting of new plants and animals into existing complex environments rather than "great batches of new species all coming in, and afterwards going out at once" were worthy of no lesser a Being.

To interpret this correspondence as indicating that special creationists were on the way to a naturalistic, that is positive, theory of evolution would be seriously to overlook its true significance. I would suggest, instead, that special creationists were intellectually incapable of such a theory and that what we have here is, at most, a cautious movement toward positivism as a general view of science. It was, after all, the growing sense that creation *must* be lawful that initiated discontent with miracle; and this was, itself, a result of the gradually developing positivism within natural history which was the entering wedge in the eventual destruction of the old episteme. It is important to note that Herschel contrasted "natural" with a "miraculous" process; and "natural," I would suggest, meant to him, not positivistic or secular causes as it would today, but rather nomothetic creationism: "natural" in the sense of being within nature, but not divorced from theological causality. Further, Herschel's pessimistic qualifications and Lyell's equivocation and the apparent insistence of both on some sort of divine initiative even in a lawful process suggests an evasion of the problem rather than a strong determination to work out a natural solution of it. A truly natural cause of new species could only mean, as Darwin realized, descent, and that could only mean transmutation—a doctrine that, at that time and for sometime afterward, repelled both Herschel and Lyell.[20]

A third option that naturalists developed, and the one that became more popular as the year of the *Origin* approached, was a reaction toward a working, almost instinctive, positivism. This was the decision, for purposes of explanatory theory-building, to leave the realm of theology and to consider the question in a purely scientific manner. This required the frank positivistic admission—whatever one's personal religious faith might be—that a unique divine act of special creation, whither by miracle or by law, was simply not a scientifically acceptable explanation since it postulated no complete and natural process that was scientifically comprehensible. As Thomas Henry Huxley remembered:

> What we were looking for, and could not find, was a hypothesis respecting the origin of known organic forms, which assumed the operation of no causes but such as could be proved to be actually at work. We wanted, not to pin our faith to that or any other speculation, but to get hold of clear and definite conceptions which could be brought face to face with facts and have their validity tested.[21]

To take this position, of course, required a commitment to the positive science that Darwin represented, but it was far from being a capitulation to transmutation. Many still rejected the descent theory but were discomfited by the lack of any plausible alternative. Whewell admitted that transmutation, which he rejected, was the only theory in the field to explain species succession. The Duke of Argyll, in his popular *The Reign of Law,* agreed that transmutation, whatever the limitations of its explanatory value as a theory, was logically superior to special creation, which offered no explanation. He, too, somewhat reluctantly, rejected it. Lyell reported paleontologist Hugh Falconer as saying in 1856 that "the opponents of the transmutation theory have never been able to propose another, so that altho' [*sic*] he himself believes in the reality of species he feels that there is a want of a counter hypothesis."[22]

J. D. Hooker, while more receptive to transmutation, was like-minded on the question of creation as a scientific issue. In his earlier work on the flora of New Zealand (1853), Hooker had followed the conventional practice of using the term creation as a synonym for "appear" or "originate," and showed little interest in the question of how species were created. But the papers of Darwin and Wallace read before the Linnean Society in 1858 on natural selection turned him around. "There must be many," he wrote in his 1860 volume on the flora of Tasmania and Australia, "who, like myself, having hitherto refrained from expressing any positive opinion, now . . . find the aspect of the question materially changed, and themselves freer to adopt such a theory as may best harmonize with the facts adduced by their own experience." As far as facts went, Hooker found it an empirical standoff between transmutation by variation and special creation. Neither theory had sufficient facts in its favor to give it a clear advantage over the other. But facts were not the only consideration: "the conclusion to which they lead, and their bearing upon collateral biological phenomena" were also important. Creationism, judged by this standard, he found sterile. It obligated science to put the means of "the origin and continuance of species"—save for "occasional variations, and their extinction by natural causes," and the "*rationale* of classification"—permanently beyond its reach, "swallowed up in the gigantic conception of a power intermittently exercised in the development, out of inorganic elements, of organisms the most bulky and complex as well as the most minute and simple." This conception was impossible: "The boldest speculator," he asserted "cannot realize the idea of a highly organized plant or animal starting into life within an area that has been the field of his own exact observation and research." Shrinking from some of the implications of transmutation, such as spontaneous generation, Hooker nonetheless adopted it for its "great organizing potential." He embraced the "newest doctrines," he wrote to William H. Harvey in 1859, "not because they are the truest, but because they do give you room to reason and reflect at present, and hopes for the future, whereas the old stick-in-the-mud doctrines . . . are all used up. They are so many stops to further enquiry; if they are admitted as truths, why there is an end of the whole matter, and it is no use hoping ever to get to any rational explanation of origin or dispersion of species—so I hate them."[23]

Huxley, who also leaned toward transmutation from intellectual necessity, told an audience of workingmen in 1859 that "the bringing into existence of an animal, at once, is a thing which is, in the nature of the case, capable of neither proof nor disproof, and is, therefore, no subject for science. . . ." If the world is governed by uniformly operating laws, he went on, the successive populations of beings "*must* have proceeded from one another in the way of progressive modification." T. V. Wollaston attributed the rise and popularity of transmutation to the "desire, which is almost inherent within us, to account for everything by physical laws, and to dispense with that constant intervention of the direct creative act which the successive races of animals and plants, such as are proved by geology to have made their appearance at distinct epochs upon this earth, would seem to require. Or," he added ambivalently, "which amounts to the same thing, it resulted through an endeavour to explain by material processes what is placed beyond their reach." Looking back into the 1840s, Lyell, who became more positivistic as time passed, but who never lost the tincture of his lawful creationism, attributed the popularity of Robert Chambers's widely discussed *Vestiges of the Natural History of Creation* to "any theory being preferred to what [Sir William Robert] Grove calls a series of miracles, a perpetual intervention of the First Cause." Later, he told Darwin that the descent theory explained much, but it did not explain everything; special creation did or could—and therein lay its weakness. What is the use of our science, he complained, if miracles may be called in at need? In the 1850s he found transmutation, doubtful as he was about it, the only "scientific hypothesis." He added,

"That new forms still rise like exhalations from the earth is no scientific theory." In April 1856, Lyell discussed the question of species with John Stuart Mill, Huxley, Hooker, and George Busk at the Philosophical Club and, on his return home, wrote in his journal that

> the belief in species as permanent, fixed & invariable, & as comprehending individuals descending from single pairs or protoplasts is growing fainter—no very clear creed to substitute . . . the successive creation of species is a perpetual series of miraculous interferences instead of the government of the organic creation by general laws.

It is interesting that, as Lyell's dilemma deepened, like others he increasingly identified special creation with miracles, which made all the easier the transition from nomothetic creation to some form of transmutation. Three years later, Huxley, writing to Lyell six months prior to the publication of the *Origin,* stressed the fruitfulness of the descent theory and the barrenness of special creation. Darwin's notion he saw as "a powerful instrument of research." He added, "Follow it out, and it will lead us somewhere."[24]

As the nonscientific nature of special creation became apparent to them, some naturalists gave it up and, publicly or privately, went over to transmutation In February 1860, A. C. Ramsey wrote to Darwin that he had "for years been using to myself the terms mutability of species, and transmutation of species, for want of better words to express the feeling (amounting to a conviction) that in Time [*sic*] species passed insensibly into each other instead of being produced by separate acts of creation. . . ." Herbert Spencer tells in his *Autobiography* that he first met transmutation in Lyell's *Principles of Geology,* but accepted it over Lyell's refutation because of a need for a natural explanation of the origin of species—a motive which Lyell admitted drew many to it. In his 1852 essay "The Development Hypothesis," Spencer attacked special creation as devoid of supporting evidence and incapable of scientific formulation. Scientists could not really believe it: its supposed credibility was merely the result of familiarity, a residue of early religious training. Many, like Huxley, must have felt that, given the facts that naturalists did have in the 1850s, something like transmutation must be true. Agassiz, on the other hand, testified to the changing orientation by complaining that some naturalists were coming "to look upon the idea of creation, that is, the manifestation of an intellectual power by material means, as a kind of bigotry. . . ."[25]

With this growing sympathy, why did Darwin bother to lash out at special creation? Because many, if not most, of those who, under the influence of positivism, had muted their creationism and had drifted away from its theological moorings, had simply embraced in its stead a second form of nescience. This was an attitude that could justify indefinitely refusing to face up to the implications of the altered idea of science that supported their change of mind: that the origins, as well as the extinction of species, must be conceptualized within the confines of a completely positive, even a materialistic, science in which God had no functional role.

Scientific nescience differed from the theologically based type discussed earlier in that, while still formally based on nomothetic creationism, in most cases it was less wed to the idea of a divine creative act. It allowed that the means by which new species were introduced was probably a continuing lawful process of which God was the author but which did not involve him directly. But the laws governing the process were unknown, would probably be unknown for a long while, and were even perhaps unknowable. "A state of doubt as to mode of formation of species [is] the sound state [of mind] at present," wrote Lyell in 1858. His appraisal of the prospects of resolving the question, which were reprinted in the *Principles* of 1850 almost verbatim from the first edition, while less pessimistic than some, could hardly be called sanguine. While Lyell did not willfully turn doubt into an obstruction of scientific growth, his slow and long-delayed public acceptance of evolution and the religious and philosophical reasons behind it suggest the rich potential for self-deception and evasive rationalization contained in nescience. The hesitancy of Lyell and others nicely complemented the tendency, commented on by Hooker, of advocates of special creation to advance hypotheses that removed the question of the origin of species from the possibility of scientific study. Richard Owen, who can stand as an example of both forms of nescience, as late as 1866 still found himself "compelled, as in 1849, to confess ignorance of the mode of operation of the natural law or secondary cause of [species] succession on the earth." Whatever Owen's secret thoughts were on the species question during this period—and his strategy seems to have been to say nothing until he could say everything—his public face, prior to the 1860s, with few lapses, was that of an antitransmutationist, and his public influence was on the side of nescience. The Duke of Argyll, who enjoyed a public reputation as a man of science, whatever his standing was in the Darwin circle, frankly doubted that the mystery of the origin of species would ever be solved. All speculations were baseless; science had no light to give. Huxley in 1859, and Asa Gray looking back from 1880, noted that many leading naturalists despaired of solving the riddle, and simply avoided the issue. A belief in the permanence of species was held with an indifference to theories about their origins. "Few naturalists," Darwin comforted Wallace—who worried about the ignoring of his 1855 essay, "On the Law which has Regulated the Introduction of New Species"—"care for anything beyond the mere description of species." Baden Powell attributed the widespread opposition to speciation theories to "a rigid—perhaps overrigid—adherence to the rule of appealing to *facts* only, and allowing . . . no *hypothesis,*" a tendency which, he warned, would destroy science if carried too far. Another cause, Powell thought, was the conclusion entertained by many that a failure of theory to date to show a precise "development from lower to higher" made all similar hypotheses improbable. And yet even he believed that an ultimate solution of the problem was "beyond the reach of positive investigation or even of human comprehension." John Phillips, apart from his theological reservations, believed that epistemological barriers shut men off from a knowledge of "created life." President Portlock told the members of the Geological Society, on the eve of the publication of Darwin's book, that the means of creating new species "still is, and probably must ever remain, a mystery." Even so, he mixed his pessimism with timid hints of evolution. It had been nescience that had driven Chambers earlier to take up the question of the origin of species.[26]

Allied to scientific nescience in its skepticism about the possibility of a knowledge of true causes was the activity, noted by Martin Rudwick, of scientists setting aside the question of the *causes* of species origination and succession and instead seeking to discover its *laws.*[27] If, however, such a quest were merely a purely descriptive enterprise, its ability to explain natural processes was limited. On the other hand, as John Herschel warned in 1845, a law that merely described the thing observed, but came to masquerade as an explanation of it, could be as definite, if more subtle, an obstruction to knowledge as a theologically grounded belief in miraculous creation. Such an approach, as Charles Bell remarked in his Bridgewater treatise on the hand, extinguished its own light: "If this be a law, there is no more to be said about it, the enquiry is terminated."[28]

The tendency to see laws as causal explanations, which was part of the Newtonian legacy, was promoted by the enthusiasm for idealism among naturalists working in the creationist episteme. A very diverse persuasion, but one finding its major spokesmen in Owen and Agassiz, biological idealism interpreted law as the manifestation of mind and the supposed fulfillment of intelligent purpose as a sufficient explanation of why a given being or structure existed.[29] "Laws, not causes," Dana told the American Association for the Advancement of Science in 1854, "are the end of true philosophy. We seek to study out the method of God's doings in nature, and enunciations of this his method or will are what is meant by the 'laws of nature.'" "A species," he wrote in his 1857 essay on that topic, " . . . is based on a specific amount or condition of concentered force defined in the act or law of creation"; and, further, "the true notion of the species is not the resulting group, but in the idea or potential element which is at the basis of every individual of the group. . . ." Phillips, in his presidential address before the Geological Society of London in 1860, stressed the limits of inductive science and the necessity of conceptualizing nature within a purposeful mental framework in order to explain and understand it. In his Rede lecture at Cambridge University the same year, he considered it

> one of the most arduous of all the enterprises of science [to] attempt to classify the various allied objects of organic and inorganic nature, according to their prevalent structures and qualities—to place them in such relation to each other as they really have—to represent, in short, the *plan* of these parts of creation—according to the leading *ideas* of which it is, or seems to us to be, the expression.

Agassiz found it one of Darwin's most serious faults that he was blind to the truth "that there runs throughout Nature unmistakable evidence of thought." Without this insight, nature was simply unintelligible. When Agassiz made the astounding statement that species do not vary "in any way," he could be understood only in the terms of his belief that only individuals existed materially in nature and that species "exist only as categories of thought in the Supreme Intelligence, but as such have a truly independent existence and are as unvarying as thought itself after it has once been expressed." Individuals might vary, but specific traits were inherited without variation. Such metaphysical subtleties were received by some with impatience. Asa Gray, for example, replied ironically that "observation shows us that they [species] do [vary]. Wherefore . . . we safely infer that the idea must have varied. . . ." Baden Powell called attention to Owen's desire in his essay *On Limbs* (1849) "to be understood as applying his conclusions solely to the *order* and *law* of succession without any attempt to assign a *cause or to trace its origin.*" And, truly, Owen's own theory of evolution by "derivation," that is, by saltation through the special birth of a modified pair of both sexes from the same mother, which he presented in the 1860s, was barren of any causal explanation other than an "intelligent Will." "A purposive route of development and change," he wrote, "of correlation and interdependence, manifesting intelligent Will, is as determinable in the succession of races as in the development and organisation of the individual. Generations do not vary accidentally, in any and every direction, but in preordained, definite and correlated courses."[30]

It was not a purely philosophic idealism itself that was the problem, of course. The laws and processes of nature might be constructively thought of as divine ideas quite as well as conceived of as ultimate facts in nature in the positivist manner. Rather, it was that tradition of religious piety which ran through Anglo-American scientific idealism, with its associations of biblicism and special creation, that encouraged the making of mysteries and inhibited research into the actual causes of phenomena. Such idealism was a natural haven for those afflicted with scientific nescience as well as for those who believed that the divine mystery of creation was forever hidden from mortal eyes. But if we could not discover the means by which God created species, at least we could discover the ends, his purposes, in that creation as they were revealed in its structure and adaptations and in the successive patterns of life. This, a disappointed Sedgwick told Darwin, was the true science which he had abandoned.[31]

Those idealistic and positivistic in their scientific sympathies alike used the concept of law as an explanation of nature. Laws were described as regulating, directing, and controlling natural phenomena.[32] In theory, it was widely recognized that the usage was metaphorical—that a law merely described the ways in which things behave. Sedgwick's three types of laws of nature—the true, the empirical, and the hypothetical—were all human constructs of varying accuracy. Whewell described laws as "somewhat metaphorical," as "rules describing the mode in which things *do* act." For Herschel, a law of nature was "a general proposition announcing, in abstract terms, a whole group of particular facts relating to the behaviour of natural agents in proposed circumstances" or "a proposition announcing that a whole class of individuals agreeing in one character agree also in another." In "common parlance," wrote John Stuart Mill, the laws of nature are "various uniformities" in nature: the "expression, Laws of Nature, *means* nothing but the uniformities which exist among natural phenomena (or, in other words, the results of induction) when reduced to their simplest expression."[33] At other times, the idea of a law seems to have been taken quite literally. Laws were seen as causal agents, or, more often, as tools of the Creator, that were products of divine will and which had an existence apart from the objects controlled. Asa Gray defined a law as "the human conception of continued and orderly Divine action." Herschel insisted that scientific formulations of law were merely recognitions of proclivities placed in matter at the Creation; God's providence worked through laws. For Agassiz, the laws of physical and living nature were "assigned to each race," were "established" by "a superior Wisdom," were as distinct creations as the phenomena they regulated, as was shown by their uniformity tying together disparate realms of nature. Owen saw laws as instruments of divine purpose, as did Chambers, Whewell, and Sedgwick.[34] Either way, the emphasis in discovering laws could easily obscure the absence of any knowledge of the actual causal processes involved.

Lastly, other naturalists, like Huxley and Gray, found the times similar to that earlier period in the development of geology described by Lyell in which theoretical indigestion had provoked a moratorium on theorizing. They turned to practical, nontheoretical fact-gathering in the good old Baconian fashion.[35]

There were then, in 1859, a minority of naturalists, some of them influential, who believed in miraculous creation; others, of shifting number, who believed in direct divine intervention in some mysterious but lawful manner to create each new species; a third group, a small minority, who had accepted the descent theory; a fourth, larger group who were moving away from a belief in direct divine intervention in favor of a natural cause, but who were either skeptical of its being found or who were engaged in a quest for laws rather than true causes; and, lastly, a group that busied itself with practical work and renounced theory altogether. All of these save the third combined willy-nilly to create a genuine obstacle in the path of the project Charles Darwin had undertaken.

How did Darwin assess this situation? His version is given in the *Origin* and in his *Autobiography*. Protesting in the latter the idea that his success resulted from evolution having been "in the air," he wrote:

> I occasionally sounded not a few naturalists, and never happened to come across a single one who seemed to doubt about the permanence of species. Even Lyell and Hooker, though they would listen with interest to me, never seemed to agree. I tried once or twice to explain to able men what I meant by natural selection, but signally failed. What I believe was strictly true is that innumerable well-observed facts were stored in the minds of naturalists, ready to take their proper places as soon as any theory which would receive them was sufficiently explained.

One's first reaction to this is that Darwin was misremembering or oversimplifying the case in order to magnify his own role in founding evolutionary biology. A more charitable and, I think, more accurate, interpretation is possible. In mitigation, it is important to remember that Lyell, on the eve of the publication of the *Origin,* agreed that most prominent "living naturalists" did believe in the permanence of species, and that Gray, as late as 1863, thought that "most naturalists believe that the origin of species is supernatural." Immutability of species was closely allied with either special creation or nescience. In the last edition of the *Origin* (1872), Darwin wrote:

> As a record of a former state of things, I have retained in the foregoing paragraphs, and elsewhere, several sentences which imply that naturalists believe in the separate creation of each species; and I have been much censured for having thus expressed myself. But undoubtedly this was the general belief when the first edition of the present work appeared. I formerly spoke to very many naturalists on the subject of evolution, and never once met with any sympathetic agreement. It is probable that some

> did then believe in evolution, but they were either silent, or expressed themselves so ambiguously that it was not easy to understand their meaning.

It is probable that when Darwin encountered scientific nescience, undoubtedly expressed in conventional creationist language or by a "reverent silence," he was misled by its evasiveness and ambiguity to the point where he tended to see it as traditional special creation. Indeed, in Owen's case he says as much.[36] The identification of immutability of species with "separate creation," given the confused use of the term creation, was not unreasonable. The immutability of species was not, of course, necessarily a theological doctrine. It was possible to treat it as a given fact of nature, much as Darwin himself treated the existence of matter, a fact that could not be explained but that on the evidence could not be doubted. This was the nescient position when stripped of theological assumptions. The only alternatives to this were special creation and transmutation.

From Darwin's viewpoint both special creation and nescience were scientific dead ends: the first asserted causes beyond conceptualization, the second no cause at all. To have acknowledged the legitimacy of either would have been to abandon science as he conceived it. Darwin, who believed that he had a causal explanation in natural selection, could not have been expected to be patient with the doubts or defeatism of scientific nescience. It is not surprising that he lumped all believers in immutability together. What this lacked in discrimination it gained in strategic force, for the nescient view and special creation agreed in wrapping the origin of species in mystery. It was to dispel that mystery that Darwin wrote his book. In so doing he was also subscribing to a new and increasingly widespread understanding of the nature and practice of science.

But turning away from special creation was not a simple matter for Darwin. It was to a later generation than his that creationist arguments became incomprehensible. For Darwin, they were still live ideas, however repugnant—an antiscience he struggled against in the *Origin of Species* and never ceased to battle against. And yet, while he broke with miraculous creation, he was not prepared to break completely with the idea of creation by law, but only with that teleological form which he encountered in those who attempted to adapt a lawful creationism to evolution. This he denounced in subsequent editions of the *Origin* as miracle, again ignoring a distinction that had great meaning for his contemporaries. But, as in the case of special creation and nescience, the result of this conflation was to underscore the need for a complete separation of biology from theology if a positive science of life were to exist.

3 Darwin and Positive Science

Darwin's relation to the positivization of biology was like that of most naturalists in his time. Conversion to that viewpoint was not a matter of philosophic reflection so much as of thoughtful day-to-day scientific practice. Darwin praised Huxley's description of scientific inquiry in his *Lectures to Workingmen* by writing, "I have often said and thought that the process of scientific discovery was identical with everyday thought, only with more care." Tutored by Charles Lyell and others, stimulated by Auguste Comte, subjected to a hundred undiscoverable influences over the years, Darwin created at Down an integration of fact and theory that has impressed astute critics ever since. But he wrote no treatise on method; his principles of science, unsystematic, aphoristic, implicit, are scattered throughout his published work, notebooks, and correspondence.[1] One must turn to these to understand the scientific sources of Darwin's attack on special creation.

Early in his career, Darwin was converted simultaneously to the view that science was an orderly system of natural causes and to a life-long aversion to theologically grounded thinking in science. The first of these had many sources, the most important being Lyell, Herschel, Sedgwick, and Humboldt. The second is more elusive. Family influences probably counted for something: his grandfather, his father, and his brother were all skeptical of orthodox Christianity and oriented toward a scientific understanding of the world.[2] His reading of Lyell's antibiblicist *Principles* must have been formative. Another, but until recently neglected, source seems to have been Auguste Comte.

To interpret nature according to "a regular system of secondary causes" alone was the heart of the positive view of science.[3] While these are Charles Lyell's words, he would not have accepted the idea that "secondary causes" alone could explain nature. Indeed, the use of the language and concepts of the older episteme in the new denotes both their dialectical relation and the ambiguity that made possible the gradual transformation of ideas about science. The traditional inductive premise that "science consists in grouping facts so that general laws or conclusions may be drawn from them" was Darwin's earliest glimpse of science as a "regular system": he learned this from Sedgwick. Leaving with Sedgwick on a geological trip through Wales in August 1831, Darwin had been immensely interested in the recent discovery of a "large worn tropical Volute shell" in a gravel pit near his home. Expecting Sedgwick to share his delight "at so wonderful a fact as a tropical shell being found near the surface in the middle of England," he was surprised that the geologist dismissed the find as the result of someone having thrown the shell away, and added, "if really embedded there it would be the greatest misfortune to geology, as it would overthrow all that we know about the superficial deposits of the midland counties." This drove home to Darwin, as his reading of science earlier had not, the nomothetic and analogical character of science. What else passed between the two men during that walking tour is unknown. Although they were separated much of the time, Sedgwick, who was talkative and pedagogical, could well have instructed young Darwin further in the philosophy of science. At any rate, passages in the *Discourse* could have been written by Darwin himself:

> Every new fact is a spark fitted to kindle in the mind of man some new train of general thought; so that the best experimenters have ever been the most successful speculators. Theory is not the idol, but the animating soul of advancing knowledge. . . . But how are we to advance one step in the way of physical truth if we shut our eyes to the evidence of analogy? . . . A good theory embodies in verbal propositions our conceptions of natural laws; and these conceptions are all based on observation, experiment, or good analogy. . . . A hypothetical spirit may do good service, provided it urge us on to make new experiments; but if we rest content with it, and, above all, if it leads us, as it has too often done, to shut our eyes against facts, and to take from nature no response but such as suits our fanatical belief of what nature ought to be, it must do deadly mischief to the causes of inductive truth.

And, lastly,

> In an advancing science, our theory may be true or false, perfect or imperfect; but as it professes to start from ascertained phenomena so must it continue to be in co-ordination with such facts as come before us during our progress, or it is good for nothing.

Darwin, in any event, got something important for the making of his positivism, perhaps more than he realized or we can know, from one of the most devout creationists in England. Even before Darwin returned from his journey on the *Beagle,* Sedgwick was convinced that Darwin would become one of the leading naturalists in Europe.[4]

The one influence that stands without serious challenge, at a time when historians are especially quick to question assertions of influence, is that of Lyell's *Principles of Geology.*[5] Lyell's postulate that, in geology (and, by extension, in all science), "*no causes whatever* have from the earliest time to which we can look back, to the present, ever acted, but those *now acting;* and that they never acted with different degrees of energy from that which they now exert" taught Darwin the lessons of an exclusive reliance on secondary causes, of economy of hypothesis, and the proper uses of analogy in scientific reasoning. The qualities that Darwin seems to have admired most in Lyell, and which he remembered in 1876 when writing his autobiography, were "clearness, caution, sound judgment and a good deal of originality." Throughout their long friendship Lyell was one of Darwin's most pressing critics and one of his most cautious supporters. In 1844 Darwin wrote to Leonard Horner, Lyell's brother-in-law,

> I cannot say how forcibly impressed I am with the infinite superiority of the Lyellian school of Geology over the continental. I always feel as if my books came half out of Lyell's brain, and that I never acknowledge this sufficiently. . . . I have always thought that the great merit of the *Principles* was that it altered the whole of one's mind, and therefore, that, when seeing a thing never seen by Lyell, one yet saw it partially through his eyes.

But, he added cryptically, "it would have been in some respects better if I had done this less." Perhaps something analogous to Sedgwick's later comment on Lyell—"he has never been able to look steadily in the face of nature except through the spectacles of an hypothesis"—ran through Darwin's mind, causing his latent Baconian prejudices to stir restlessly. Or, more likely, it was an awareness of his divergence from Lyell's antitransmutationism and the need for proving his independence; for, whatever the attractiveness of Lyell's thought, its nescience made it a flawed tool in Darwin's hand. He had to make his own.[6]

Other intellectual influences in shaping Darwin's ideas about science are more problematical. During his last year at Cambridge, he read Alexander von Humboldt's *Personal Narrative* and John Herschel's *A Preliminary Discourse on the Study of Natural Philosophy.* These books, he wrote, "stirred up in me a burning zeal to add even the most humble contribution to the noble structure of Natural Science. No one or a dozen other books influenced me nearly so much as these two." Darwin read Herschel's *Discourse* again in 1838 at the time of his momentous encounter with Thomas Malthus's essay on population and also, at that time, worked through William Whewell's *History of the Inductive Sciences.* Great claims have been made for this reading by those interested in tying Darwin's accomplishments to the rules of contemporary philosophy of science. Michael Ruse, who makes a most plausible case for this view, admits, however, that the Newtonian scientific ideal embodied in Herschel's book was general in Britain in the 1830s.[7] In truth, Darwin would have been hard put to have avoided it. Whether "being scientific" meant being like Herschel or whether it meant being like Lyell, it meant being positivistic, and this was the model that young Darwin chose. But, whatever Herschel's direct and formal influence on Darwin might have been, he was undoubtedly one of his inspirations. Herschel's opinion, good and bad, mattered. Darwin was delighted, in 1838, when he heard that the famous astronomer had told Lyell that the origin of species involved "intermediate causes"; and he was crushed years later when he heard rumor that Herschel had dismissed natural selection as "the law of higgledy-piggledy." "What this exactly means," Darwin lamented to Lyell, "I do not know, but it is evidently very contemptuous. If true this is a great blow and discouragement."[8]

Alexander von Humboldt's part in sending Darwin to the tropics is well known. Overwhelmed by the beauties of nature as described in the *Narrative,* Darwin and his Cambridge teacher and friend J. S. Henslow planned an expedition to Tenerife: "I hope you continue to fan your Canary ardor," Darwin wrote in 1831; "I read & reread Humboldt, do you do the same, & I am sure nothing will prevent us seeing the great Dragon Tree." It was not with Henslow, but in the *Beagle* that he was to see the glories written of by the great traveler. "I now almost adore him," Darwin wrote from Brazil. His diary during the early months of the voyage was filled with confirmations of his expectations. Humboldt "like another sun illumines everything I behold." He had sour days: "As a sultan in a Seraglio I am becoming quite hardened to beauty. It is wearisome to be in a fresh rapture at every turn of the road. And as I have said before, you must be that or nothing." Only a few days later, however, his raptures were back to normal. After Darwin left Brazil, Humboldt virtually vanished from his diary, but not from his mind or memory. Recent writers have emphasized Darwin's debt to Humboldt's conception of natural history, of ecology, populations, and temporal change. He encouraged Wallace to publish his own travels to inspire others to study natural history; and, in 1881, praised Humboldt to Hooker as a true savant: his "omniscience"—although his geology was "funny stuff"—and his example in inspiring "a grand progeny of scientific travellers, who, taken together, have done much for science" were the sources of his merit. Darwin never lost his love of beauty and of nature; and this was not unimportant in explaining his conversion to positive science. Humboldt taught him, and others, the extent to which a naturalist could satisfy the needs of the spirit through his work.[9]

It was on the *Beagle* voyage (1831–36) that the idea of positive science as a mode of knowledge came together for Darwin. His geological work, particularly in South America, gave him practical experience in scientific reasoning (on Lyell's principles), and was capped by his theory of the development of coral reefs owing to subsidence of the ocean floor which was his first successful major scientific explanation.[10]

Early in the voyage, Darwin feared he lacked the knowledge to make correct observations, and that what he did would not be of any interest to others. Undoubtedly, this increased his determination to do well, and this in turn stimulated both his observational and theoretical gifts. In the absence of an adequate geological library, he was forced to draw his own conclusions, "[and] most gloriously ridiculous ones they are," he wrote in 1834 to Henslow, who had been advising him. "I sometimes fancy I shall persuade myself there are no such things as mountains, which would be a very original discovery to make in Tierra del Fuego." Reconstructing the unknown fascinated him: "The limit of man's knowledge in every subject," he wrote in his diary in 1834, "possesses a high interest, which is perhaps increased by its close neighbourhood to the realms of imagination." Repeatedly, he sought and found verification of Lyell's doctrines. He offered Henslow elaborate inferential reconstructions of geological history which he did not expect him to believe, but in which he had growing confidence. Unlike his occasional remarks on species, the geological theories contained no theological elements and both presumed and stressed the uniformity of law and secondary causes. In speculating on how the scattered small islands of the Chronos archipelago in Chile were populated with small mammals, Darwin appealed only to "a succession of chances," that is, to the transportation of individuals or to "changes of level" of the land. Although still willing to invoke the creation of new species, he was not willing to populate every island in that manner. Perhaps we have here signs of a growing sense of the inutility of the idea of special creation and a corresponding awareness of the necessity of explanations involving observable causes. However this may be, the voyage revealed Darwin increasingly impatient with conventional biblicist analogies and inhibitions. After recounting the destruction of life on the pampas during the drought of the 1820s, he asked, doubtless thinking of Buckland's diluvialism,

> What would be the opinion of a geologist, viewing such an enormous collection of bones, of all kinds of animals and of all ages thus embedded in one thick earthy mass [i.e., in river deposits]? Would he not attribute

> it to a flood having swept over the surface of the land, rather than to the common order of things?

Like many a protopositivist before him, he was amused by theological explanations of fossil shells on mountain tops: "born by Nature" or "God made them."[11]

The absence of cosmological and theological speculation and explanation in Darwin's geological writings, to be sure, is to a degree owing to the nature of the medium. It means not so much that he was consciously casting off creationist modes of thought as that he was following the descriptive and explanatory practices of others in the science which, as in the case of the Geological Society of London, continued the antispeculative prejudices of English Baconianism. But this is the point: Darwin did his initial writing in science and his premier theoretical work in a disciplinary atmosphere that was conducive to developing a positivist attitude. Unlike Asa Gray, who matured in botany when a taxonomic creationism reigned and who hated the positive outlook, Darwin early accepted it as the only viable approach to resolving scientific questions.[12]

By 1838 this orientation in science was characteristic of his biological writing as well. Sometime between the spring of 1836 and the fall of 1838, when he was probably irreversibly committed to a belief in speciation by descent with modification (with whatever lingering doubts), Darwin realized that creationist explanations in science were useless. Once this had been decided, transmutation was left as virtually the only conceivable means of species succession, certainly the only one that could be investigated. Hence, as Howard Gruber has pointed out, the descent theory provided Darwin with a system of laws organizing life—or, to be more accurate, with the problem of identifying those laws—long before he had any idea of its explanation.[13] When Darwin began to consider the problem of species extinction, succession and divergence, he did so as an evolutionist because he had first become a positivist, and only later did he find the theory to validate his conviction. The well-known evidence gathered on the *Beagle* voyage—the similarity of living and fossil species in South America, the distribution of species on the Galapagos Islands, and so forth—became evidence, and its explanation a problem, only because Darwin had grown accustomed to positivistic explanations as a result of bending his mind to geology. These things were not problems for creationists. When Darwin responded so enthusiastically to Sir John Herschel's belief that new species were introduced by a lawful process—"intermediate causes—Hurrah!"—he failed to appreciate that creation by law, while the bridge over which many passed to transmutation, could be as obfuscating as creation by miracle, perhaps more so, because of its apparently scientific formulation.

An acceptance of the positivist idea that nature is best explained as a system of natural law and natural causes, however, did not complete Darwin's passage to positivism. It was necessary to make explicit what was implicit in that view: the influence of theology on science had to be rejected.

The presence of a biblicist mode of thought in scientific discourse was a major element in the conflict between science and religion in the nineteenth century. The point raised by biblicism was not whether literalism in biblical interpretation should survive, nor was it whether science contradicted the Bible. Although these were important questions, they were not central and, in any case, were virtually dead issues among men of science. Theologians could deal with these questions. They had faced Augustine and Calvin in their day just as they now faced the clergy of the nineteenth century. The point of conflict, rather, was the intellectual autonomy of science. The real issue was not the validity of scriptural geology but the continuing use of biblical themes and images in scientific thought quite apart from a biblical literalism. Few scientists, by mid century, any longer believed in the literal historicity of Noah's flood—at least the majority had given up looking for geological proofs of it—but many would not give up some sort of historical flood. Few believed in the biblical Adam and Eve, but almost everyone talked of "a single pair" in discussing the origin of species and saw man as a unique creation. Few believed in miracles, but many still spoke of creation in ambiguous terms. Few any longer thought that Genesis told the actual story of the world's creation, but some still tried to find "periods" of geological activity roughly parallel to Moses' six days and their respective events. To the positivist these biblical images were repugnant vestiges of an older and discredited science, one that had too many theological loyalties and intellectual inhibitions. To be sure, scientists often had many facts to support their biblicist theories, but it was the near-prescriptive and a priori nature of the theories that was the problem. In a real sense, the answers were predetermined by the old questions and their assumptions, and new questions were not asked. Lyell made the point clearly to the Duke of Argyll in 1869:

> The assumption of special creation would make the origin of all things simple; but if once a naturalist taking all the geological evidence into account inclines to the opposite view, that of transmutation, as more probable, we are led to speculate on the means in which present instincts, habits, structure, and colour came to be gradually acquired by each species. . . . The transmutationist may reasonably hope that we shall get to understand more and more the working of those forces by which nature brings about changes in the organic and inorganic world. . . . [The study of secondary causes] will help to explain many perplexing phenomena which those who are satisfied with special creation would never bring to light.[14]

The obstructing and distracting influence of theological and biblicist presuppositions in science had been virtually the first lesson that Lyell had taught in the *Principles of Geology*. The biblical flood, the traditional short chronology for the age of the earth, the supposed recent date of the creation of contemporary species of animals and plants, all the contrived attempts to make geology compatible with Genesis, had served, he thought, only to prevent accurate observation and the formulation of sound theory.[15] Closely identifying the theological influence on science with biblical geology, Lyell attacked both with vigor and sarcasm. Always careful of the religious sentiments of his readers, he feinted with a denunciation of the cosmogonies of Hinduism as "puerile conceits and monstrous absurdities" embedded in a "pretended revelation," and then went on, subtly:

> The superstitions of a savage tribe are transmitted through all the progressive stages of society, till they exert a powerful influence on the mind of the philosopher. He may find, in the monuments of former changes on the earth's surface, an apparent conformation of tenets handed down through successive generations, from the rude hunter, whose terrified imagination drew a false picture of those awful visitations of floods and earthquakes, whereby the whole earth as known to him was simultaneously devastated.

For Lyell, scientific diluvialism and catastrophism clearly had their primal origins in priestcraft and superstition. Their continued theoretical employment, even by geologists as skilled as Elie de Beaumont, did not validate them. The biblicism that underlay even theories such as his was an "incubus" that, in Lyell's opinion, had done great harm to the science.[16]

The continuing influence of biblicism in science was not merely a personal phobia of Lyell. While even the most orthodox naturalists condemned crude biblical literalism, they would not turn biblicism loose. Dana took British diluvialist William Buckland to task before the American Association for the Advancement of Science in 1855 for doing geology out of the Bible rather than out of the rocks. And yet, as late as 1880, he was himself still reconciling geology with Genesis. Sedgwick likewise rejected scriptural geology as scientifically ignorant. The unfortunate Dean of York who thought that the world might well have been created in six days fell under his condemnation in 1844, as did the more experienced Hugh Miller in 1850. But, like Dana, Sedgwick could not resist the urge to show that Moses was, in some sense, right after all. In 1858 he wrote to a friend that schemes of reconciliation were wrong because they were premature. In fifty years, he thought, we would be in better shape to compare geology and the Bible; he was confident that reconciliation would be achieved. The biblicist spirit was prevalent and enduring. Owen, in his 1858 presidential address to the

British Association, found it still worth his while to oppose the vestiges of a belief in the Mosaic flood and the subsequent spread of all animal life from a single center in southwest Asia. "The cultivators of our science," Lyell told the members of the Geological Society of London in 1850,

> may be ready to grant the most indefinite duration to each successive geological epoch, yet they may still unconsciously derive a love of cataclysms and catastrophes, and faith in a primeval chaos out of which the present order of things was evolved, from an hereditary creed, not founded on facts, or strict inductive reasoning on natural phenomena.

J. B. Stallo, author of *The Concepts and Theories of Modern Physics* (1881), found biblicism even in the nebular hypothesis of Laplace: the assumption that the world "in the beginning" must have been "without form and void."[17]

Not all instances of biblicism, however, involved a direct theological influence. In his New Zealand flora, Hooker adopted the view that species were immutable (save for local variation) and each descended from a single pair. This was not necessarily his belief, but a methodological postulate to make classification possible. This, one of the most common instances of biblicism in natural history, disturbed Asa Gray, who suggested substituting for "single pair" the phrase "common stock," which could better accord with the facts and requirements of nature. Hooker, however, continued the usage. In the flora of Tasmania (1860) he considered that the geographical distribution of species from a common center was "all but conclusive evidence in favour of the hypothesis of similar forms having had but one parent, or pair of parents." Hooker, it should be noted, had by this time been a Darwinian for two years. The doctrine of centers of creation, that each species was thought to have originated as a single pair in a given locality, was, with its aura of Eden, historically an instance of biblicism, if not logically so, as Hooker's usage, and Wollaston's also, show. It could, and did, promote work in understanding the nature of natural barriers and was easily incorporated into Darwin's theory. Even so, there was uneasiness even about "single pairs," and not all biblicism was so useful.[18]

A more severe attitude toward biblical themes in science was voiced by Huxley in 1859 when he warned an audience of workingmen to beware of "bigoted orthodoxy" and the clergy, its instruments. Looking back in 1893, he remembered how frequently during his youthful scientific work he had encountered a "theory barrier with its comminatory notice-board—'No thoroughfare. By order, Moses.'" This was a manifestation, not of Christianity, but of "the dominant ecclesiasticism of my early days, which, as I believe, without warrant from the Bible itself, thrust the book in my way."

Similarly, Hooker expressed to Gray his resentment of "parsons" in science, naming Whewell, Baden Powell, Sedgwick, and Buckland. He wrote to Harvey in 1860, "I hold Whately and Sedgwick to be as really *ignorant* of the fundaments of Natural History as I am of Church History or you of fluxons." It was felt that such men were caught in a double loyalty, and could not be relied on in science. Antagonism against biblicism had reached such a point by the late 1850s that Agassiz suggested that fear of the wrath of the positivists was actually leading some naturalists with strong theological convictions to conform to the new science against their true judgment. One might well doubt the reality of such a reign of terror, but there can be no question of the hostility felt by the antibiblicists.[19]

Lyell and the others, of course, were not enemies of theology as such. Lyell's aim, like that of Bacon and Galileo before him, was to protect science from the inhibitions and misdirections of theology. This he did, not in order to harm religion, but in order to serve science. His lesson was not lost on Darwin, but his disciple was to go far beyond him. In one of his early critical writings, Lyell asserted that the study of fossils showed that all beings and all ages in the earth's history were part of a single plan proceeding from "the same Author" and "One Mind." He revealed himself as anxious to marry geology to the conventionally popular religious apologetics of Bishop Joseph Butler, and praised the "wisdom, justice and benevolence of the common Parent and Preserver" of all. This basic theological conviction never left him. Thirty years later, when he was struggling with the species question, one of the problems he set himself was "whether the creative power in which I conceive design & supreme intelligence is manifested, is or is not governed by the ordinary laws of continuous generative succession."[20] Lyell's care not to offend churchmen was not merely a matter of tactics (although it was that), but also reflected a genuine concern for theology in its proper sphere. He was determined to go no farther in what all recognized to be a potentially antichristian direction than was necessary to establish sound geology. It would be a serious error to interpret his antibiblicism and his light bantering letters about "orthodoxy" as expressing hostility to religion or to Christianity. It was theological interference with science, and not doctrine, that concerned him.[21] Was Lyell "disingenuous" when he stressed to the bishop of Llandaff the extent to which his geological system was compatible with a biblical history?[22] Or, rather, was he not struggling with a real dilemma: the positivism of his science and its methodological requirements, versus a desire to maintain a viable theism containing the possibility of some sort of divine activity in the world? Like Darwin and others, Lyell was caught in a profound shift of mind, one that was not easily executed. While fostering the growth of a uniform, naturalistic view of nature that eventually would leave

God with nothing to do, he could, at the same time, recommend books which reconciled "God's free will to the immutable laws of Nature."[23]

Despite his attacks on biblicism in science, Lyell ended the *Principles* on the same note of mystery-mongering that Darwin later condemned. After rejecting the attempts of Buckland and others to bring geology more into accord with Genesis, Lyell wrote:

> For our own part, we have always considered the flood, if we are required to admit its universality in the strictest sense of the term, as a preternatural event far beyond the reach of philosophical inquiry, whether as to the secondary causes employed to produce it, or the effects most likely to result from it.

The abandonment of his own principles of uniform causality and analogical reasoning, and the sophistry that followed from it (that the "preternatural" flood waters might have engulfed the delicate volcanic cones of the Auvergne without disturbing them!), was of a piece with his finally drawing a veil of pious nescience over the origins of both the earth and species.[24] The seeds of positivism were planted in Lyell's mind but they sprouted slowly and never as luxuriantly as in the case of his young colleague in Kent.

In 1859 the Duke of Argyll exemplified the traditional nonpositivistic view of science by saying that natural laws could not be the bases of really full explanations. They told only "the rules," the processes which nature follows. They did not tell "the how or the why." These questions, the ultimate ones, were beyond science and must be answered by religion. Science, Argyll asserted, was only "the vestibule of the Church." While Lyell, perhaps, would not have dissented from the strict letter of this, in his more positive moods—which had become more regular and lasting during the years since 1830—he would certainly have abominated its spirit.[25] Darwin, of course, firmly rejected the slightest hint of a dependence of science on religion but, like Lyell, his main intention was to promote science, not to attack theology. And yet, the result was the same: revelation, at worst a source of mischief, at best irrelevant, had no role to play in describing or explaining the world of nature. He attributed much of the opposition to his theories to theological prejudice, and found a lingering biblicism even in Lyell.[26]

Louis Agassiz stood second to no man in his opposition to sectarian religious interference with science. What, then, separated him from Lyell or, better, from Darwin? It was the theological ground of his view of science. While Agassiz is admittedly an extreme example of the theological scientist, he was not unique, and his very atypicality makes him useful for purposes of contrast. In his Graham lectures (1862), Agassiz defined the study of nature in a way that clearly indicated the problem of theology in science. The "one great object" of the study of nature, he said, "is to trace the connection between all created beings, to discover, if possible, the plan according to which they have been created, and to search out their relation to the great Author." This program presupposed creation, a plan and purpose (which may not be discoverable by man), and channeled the scientist's theoretical efforts into metaphysics and theology. "We must," he had written earlier in criticism of Darwin, "look to the original power that imparted life to the first being for the origin of all other beings, however mysterious and inaccessible the modes by which all this diversity has been produced may remain for us. The production of a plausible explanation is no explanation at all, if it does not cover the whole ground." The "whole ground," for Agassiz, included theology.

That Agassiz's science was free of theological influence because he did not base it on the tenets of a conventional religious creed would have been considered an illusion by any positivist. There was nothing in the theological perspective that required scientists to be shallow in their views, of course. As the directors of the Graham lectures pointed out:

> The man of science . . . will trace the hand of God, His skill, goodness, and wisdom, at every step of progress in his studies. He will discover more minutely that sublime wisdom which has made the wonderful organisms of plants and animals, and arranged the immensity of nature.

Such a persuasion could and did countenance scientific research of great complexity. Why then the positivist objection? Because it drew a veil of mystery over the processes of nature. "Contrivance" concealed the operation of natural selection; "creation" inhibited inquiry into the causes of species succession, geographical distribution, and other subjects. The theological scientist could describe the operation of the laws of nature, but he could not explain it. He could marvel at the ingenuity of it, but his ability to understand the interrelated self-sustaining operation was lessened by his presuppositions which encouraged him to substitute praise for knowledge. If the positivist was arrogant, his arrogance was more fruitful in the long run. Agassiz closed his lectures with these words:

> Here in the animal kingdom we see [the ingenious and economical contrivance of the creator] illustrated to an extent which the best-trained mind can hardly follow, showing how far beyond our comprehensions are the wonderful works of nature. Even though we can make ourselves conscious that they are built by mind, and that it has pleased the Maker of all things to give us a spark of that life which makes us to be His children, formed in His image, the evidence is nowhere stronger than in the fact

> that our mind is capable of studying those works to a limit which approaches to a comprehension of their wonderful relation to one another.

Agassiz described, classified, and rushed on to teach a theological lesson. He showed great reluctance to explain natural history fully in natural terms; indeed, his epistemic assumptions precluded such an explanation. His most positivistic effort, the ice age theory, but cleared the earth for a new outpouring of creative energy. According to Gray, it was Darwin's proficiency in providing "good physical or natural explanations" that most annoyed Agassiz.[27]

The naturalists of the old episteme, while developing in their practice a protopositivism as manifested in their preference for nomothetic creation and in their purely naturalistic work in geology (in which Agassiz, of course, was a leader), were, at bottom, still engaged in a process of naming such as Foucault found in the natural history of the eighteenth century: an "essential nomination," a "transition from the visible structure to the taxonomic character." Classifying was the genius of this science and the divine paradigm, incarnated in all true classification, was its ultimate source of meaning.[28]

Those who argue that there was no real warfare between science and religion in the nineteenth century ignore the presence of these two sciences. The old science was theologically grounded; the new was positive. The old had reached the limits of its development.[29] The new was asking questions that the old could neither frame nor answer. The new had to break with theology, or render it a neutral factor in its understanding of the cosmos, in order to construct a science that could answer questions about nature in methodologically uniform terms. Uniformity of law, of operation, of method were its watchwords. The old science invoked divine will as an explanation of the unknown; the new postulated yet-to-be-discovered laws. The one inhibited growth because such mysteries were unlikely ever to be clarified; the other held open the hope that they would be.

During the years when young Darwin was throwing off the constraints of theological science, he had a brief, indirect, but significant encounter with the thought of Auguste Comte. Despite his later contempt for Comtian positivism, this fortuitous meeting seems to have been the same sort of pivotal moment as was his reading of Malthus close to the same time. In September of 1838, Darwin wrote to Lyell:

> By the way, have you read the article, in the "Edinburgh Review," on M. Comte, "Cours de la Philosophie" (or some such title)? It is capital; there are some fine sentences about the very essence of science being prediction, which reminded me of "its law being progress."

What was in this review? In addition to his attack on teleology and final causes, Comte's position that "all real science stands in radical and necessary opposition to all theology," in view of what followed, must have caught Darwin's eye despite the reviewer's condemnation. Comte's system of three stages in the development of each science—theological, metaphysical, and positive—was acknowledged to be generally true by his critic and quoted at length. Interestingly, Comte's name then began to turn up in Darwin's notebooks:

> M. Le Comte's [*sic*] idea of theological state of science. grand *idea:* as before having analogy to guide one to conclusion that any one fact was connected with law—as soon as any enquiry commenced. for instance probably such a thing as thunder would be placed to the Will of God.—Zoology itself is now purely theological.

And, elsewhere, "M. le Comte argues against all contrivance—it is what my views tend to."[30]

It would not do to put too much emphasis on this incident. The origins of Darwin's positivism are deep-rooted and complex. The timing, rather than the content, may have been the important thing. But, however one assesses it, it is clear that Comte's "grand *idea*" gave Darwin a concept when he needed it and one which stayed with him. In 1861 he told Lyell that most people's thought on species "is still . . . under its theological phase of development," and that "I must think that such views of Asa Gray and Herschel merely show that the subject in their minds is in Comte's theological stage of science."[31] Ironically, his own thought did not completely escape that stage, but there is no question that it did so on the species question and other scientific issues. In such inquiries, theology became merely redundant. Positive science, conceived as a system of empirically verifiable facts and processes and the theories linking them, required the radical desacralization of nature. There could be no out-of-bounds signs and, despite its emphasis on a rational correspondence of the divine mind and man's, the old science did erect such barriers. God's initial processes of creation and his ultimate purposes were unknowable. Too much of the content of the old science was the result of intuition that was in principle unverifiable, either directly or indirectly.

What, then, were the reigning principles of Darwin's view of science when he wrote the *Origin?* He assumed, like most positivists, a system of natural causes operating according to uniform laws of nature. Neither unique causes nor absolute chance could be predicated to exist in the universe. But, unlike some, he did not turn this assumption into a system of materialist metaphysics. In point of fact, Darwin held a dual notion of natural law. In a higher sense, as we shall see, the laws of nature were God's creation, expressions of his will that regulated the world processes. Ideally, these were also the objects of scientific knowledge. Scientific knowledge, however, was made up of man's relative and fallible understanding of these laws. Hence, the working laws of scientific theory were not real but rather constructs that hopefully approached reality.[32] In the *Origin,* therefore, he defined "nature" as "only the aggregate action and product of many natural laws" and "law" as "the sequence of events *as ascertained by us*" (emphasis added).[33]

Like most of his contemporaries, in practice Darwin used the term law in various ways. Sometimes he spoke metaphorically of laws that stand in the place of unknown causes, as, for example, laws that "govern" the sequence of life or variability, the "law of common embryonic resemblance" which larvae "obey," and "the laws governing inheritance [which] are quite unknown." Often he used the term as a generalization of regularity in phenomena, for example, the "laws" and "rules" of hybridity. He also spoke metaphysically, for example of the "laws impressed on matter by the Creator," or of a general law leading to the progress of all beings. This last was not, of course, an advocacy of a "fixed law of development" which required the uniform evolution of all species, a concept which he denounced. These uses, save the metaphysical, which falls into a special theological category to be taken up later, conform to his definition of law as an operational feature of science and as a human construct.[34]

Darwin's view of natural law as tentative in science carried over into an instrumentalist conception of theory. "In scientific investigations," he wrote in *The Variation of Animals and Plants under Domestication* (1868), "it is permitted to invent any hypothesis, and if it explains various large and independent classes of facts it rises to the rank of a well-grounded theory." Here, "explain" seems to have meant to relate facts by a system of theoretically, if not yet actually verifiable causes. Unique or preternatural causes, those which necessarily could not be investigated, were ruled out on actualistic principles.[35]

Absolute chance or spontaneity was as opposed to the conception of a lawful and predictive science as the unique or preternatural cause. Although Darwin used the term chance, it often meant only "our ignorance of the cause." Sometimes he used it in the conventional sense of purposelessness, although by that word Darwin usually meant without a place in the structure of nature, rather than without an intelligent cause and goal, which is what most Victorians meant by it—and which he did also on occasion. Hence, in the *Descent of Man* he wrote both of "the grand sequence of events" in living nature not being "the result of blind chance," that is, without a higher purpose and also, "it is incredible that all this [that is, sexual characteristics] should be purposeless," meaning without a function in nature. At other times, he used chance in the sense of random, fortuitous opportunity:

> It is improbable that the unions of quadrupeds in a state of nature should be left to mere chance. It is much more probable that the females are allured or excited by particular males, who possess certain characters in a higher degree than other males. . . .

And conversely, in relation to natural selection, he wrote, "I had always perceived that . . . the preservation of even highly-beneficial variations would depend to a certain extent on chance," that is, on varying, haphazard, and coincidental circumstances. But far from building his doctrine on absolute chance, Darwin insisted that all phenomena are governed by laws and so are potentially subject to scientific study and explanation. At no time did he see the world as the result of fortuity in a metaphysical sense. This was contrary to both his science and, as will be seen, his theology.[36]

The assumption of a uniformity of lawful causality, by which absolute chance was ruled out, made possible the employment of analogy, which was the real backbone of theory-building. Analogy had enabled Lyell to read the history of the earth:

> [The] immutable constancy [of natural law] alone can enable us to reason from analogy, by the strict rules of induction, respecting the events of former ages, or, by a comparison of the state of things at two distinct geological epochs, to arrive at the knowledge of general principles in the economy of our terrestrial system.

By this means the present illuminated the past and "existing analogies" kept a check on wild speculation regarding geological agents. It had been a realization of the analogous processes linking "the ancient and modern state of our planet" which French geologists had established earlier that "slowly and insensibly" had drawn men's minds away from "imaginary pictures of catastrophies and chaotic confusion." If, Lyell wrote,

> instead of inverting the natural order of inquiry, we cautiously proceed in our investigations, from the known to the unknown, and begin by studying the most modern periods of the earth's history, attempting afterwards to decipher the monuments of more ancient changes, we can never so far lose sight of analogy, as to suspect that we have arrived at a new system, governed by different physical laws.

Analogy was the sine qua non of a positive geology. It was implicit in the practice of uniformitarian and catastrophist alike, and only waiting to be formulated and raised to an absolute principle.[37]

While recognizing in his mature work that analogy might be a "deceitful guide," "an unsafe support," and a source of fanciful opinions, Darwin early realized its importance in theory-building and other forms of scientific reasoning. Commenting on William Whewell's remark, in 1839, that "if we cannot reason from the analogies of the existing to the events of the past world, we have no foundation for our Science," young Darwin wrote enthusiastically: "but experience has shown we can & that analogy is a sure guide & my theory explains why it is a sure guide."[38] His use of analogy became one of the most controversial features of his method. In addition to the conventional employment of analogy to connect past and present, Darwin used it heuristically to suggest hypothetical processes and causes. The many arguments from domestic breeding to illuminate natural selection; the evolution of the eye and other complex organs; the hypothetical single ancestor of all living forms; and other examples, filled the pages of his work and, despite the sanction for such thinking in science given by the philosophers Herschel, Whewell, and Mill, were eagerly seized on by his critics.[39] He used a "great mass of analogous facts" to battle Wallace on the theory of birds' nests and expressed surprise, in 1856, at Lyell's "thinking it immaterial whether species are absolute or not." How natural history would be transformed, he exclaimed, "whenever it is proved that all species are produced by generation, by laws of change." Then analogy might fill those troublesome gaps in geological formations. When his vicar at Down, the Reverend J. Brodie Innes, reported seeing a cross between a cow and a red deer at a stock show, Darwin discounted the wonder as "opposed to analogy." Analogy, then, both opened up biology and, at the same time, provided it with a necessary structure of probability.[40]

Equally important with the concept of the uniformity of nature and its corollaries was the principle that science should properly deal only with theoretical explanations based on actually or potentially observable or verifiable causes. Darwin saw clear parallels between Lyell's geology and his own theory of natural selection in this regard. Descent with modification, and migration, was the only *vera causa* capable of explaining species formation and distribution without resorting to purely hypothetical explanations, mysteries, or miracles. In the *Origin,* as in the *Principles,* we find a sustained attempt to avoid ad hoc theories that will "explain" the phenomena and, instead, to construct theories that will organize all the facts within hypotheses based on patterns of causes known to exist and whose operations are observable. In Darwin's view, such theories, whatever their flaws might be, promoted research; the others retarded it.[41]

This attitude in science, which was vital to the development of positivism, came to be called "actualism" in the last century, and currently goes under that name. W. F. Cannon's definition of the principle as applied to geology would win agreement from most: "the tendency to refer geological events and monuments to causes similar in nature, quantity, and intensity . . . to those observable in principle at present"; an actualist, he goes on, is not "one who merely is willing to accept some current causes, but one who favours them." The actualist favored them because without the mutual support of actualism, analogy, and the uniformity of the laws of nature, science would not have been possible—or not possible, at least, in the way that positivists wished to practice it. As a general principle, this form of actualism was the common sense of science. No one would have disputed it. What was disputed was the "absolute actualism," to use Cannon's term once more, adopted by Lyell and generally known as uniformitarianism. Those who held this position insisted not only that modern causes be postulated as the only causes operative in past time, but that the degree of their operation be identical with that presently observed.[42]

This doctrine became the basic point of contention between Lyell and his catastrophist opponents. The latter founded their objections on facts: there were certain geological remains that simply could not be accounted for by the present-day intensity of geological operations. Roderick Murchison, for example, argued that only two known forces could have produced the geology of the plains of Germany: glacial action or water action. Glaciers were clearly impossible. Water must have been the agent. But modern water action, he claimed, was incapable of producing the observed effect—the movement of boulders over great distances. Murchison therefore invoked "waves of translation" caused by sudden elevations of the land to account for the erratic drift. This uplift was not slight or gradual—such as Lyell could sanction—but "successive sudden upcasts which threw off great devastating and erosive waves." Murchison's wave of translation was a hypothetical creation admittedly lacking modern empirical verification. It was based on experiments and mathematical calculations which showed that only a solid wall of water moving at a speed of twenty-five to thirty miles per hour could have done the work. In this Murchison, unlike some other catastrophists, clearly had no Mosaic axe to grind despite the "flood" motif. His reasoning was purely scientific and not unlike Darwin's in his use of analogy and his dependence on partial corroboration of events that could not be witnessed. It differed, however, in that Murchison, unlike Lyell, was invoking causes for which he had no evidence other than their presumed effects. Darwin, on the other hand, and like Lyell, tried to construct causes from known or at least highly probable factors: variation, heredity, population pressure, and a struggle for survival. After all was done, Murchison had no independent evidence that the earth had ever behaved as he claimed; it was as if Darwin had used only pigeon breeding and fossils to argue evolution. But Murchison did not stand alone. Whewell, who also opposed Lyell's extreme actualism without questioning the actualistic principle itself, argued that "the great mass of northern drift . . . is an irresistible evidence of paroxysmal action," and that "the distribution of the northern drift belongs to a period when other causes operated than those which are now in action."[43]

Lyell, in the judgment of many of his peers, flew in the face of facts. But we can understand this when we realize that sound method was more important in his view than facts. Indeed, method could change the facts. "It is not the magnitude of the effects, however gigantic their proportions," he told the Geological Society in 1850, "which can inform us in the slightest degree whether the operation was sudden or gradual, insensible or paroxysmal." The actualistic principle, unchecked by uniformitarian caution, had led the catastrophists to beg the question in assuming that large effects meant large and sudden causes and those causes ones not now in operation. It must be shown, he said, that present causes are incapable of executing the changes in question.[44] For Lyell, nonuniformitarianism meant not only hypothetical causes, but an obstructing nescience as well. Like many nomothetic creationists, nonuniformitarian geological actualists admitted the operation of natural causes, but these were sometimes causes that, in nature and operation, were unknown and unknowable. Not all catastrophists, to be sure, were so unimaginative as to be without a theory. But, in Lyell's view, imagination was not what was needed: a true method was what a positive geology required.

The principles of Lyell's uniformitarian method were neither historical nor empirical in their origin. They were the logical corollaries of analogical reasoning. Martin Rudwick has pointed out their theological sanction in Lyell's thought: that it was an idea more worthy of God to have designed an interdependency of nature to insure ecological balance and uniformity. This was the well-known steady-state feature of Lyell's geology. It was a system held in equilibrium in which directional change was inconceivable.[45] But its origins were not entirely theological, nor even primarily so. At bottom, Lyell's method was deeply committed to positivism. His first step was to get rid of the biblicist intellectual heritage. He then established an approach to analyzing geological phenomena based entirely on actualistic and uniformitarian ideas that stressed economy of hypothesis and the employment of observable causes. He structured his science on principles that were almost mathematical in their abstraction, but which, nonetheless, gave geologists a workable set of assumptions. "In order to confine ourselves within the strict limits of analogy," he wrote,

> we shall assume, 1st, that the proportion of dry land to sea continues always the same. 2dly, That the volume of land rising above the level of the sea, is a constant quantity; and not only that its mean, but that its extreme height, are only liable to trifling variations. 3dly, that both the mean and extreme depth of the sea are equal at every epoch; and 4thly, It will be consistent, with due caution, to assume, that the grouping together of the land in great continents is a necessary part of the economy of nature. . . .

His book, he wrote to Murchison in 1829, will "endeavor to establish the *principle of reasoning* in the science . . . that *no causes whatever* have from the earliest time to which we can look back, to the present, ever acted, but those *now acting;* and that they never acted with different degrees of energy from that which they now exert." Past and present could only be connected by strict analogy; without it, one had not science, but fancy.

Lyell denied that his assumptions meant an a priori exclusion of "paroxysms and crises." He only urged that they should not be assumed, a priori. We know, he argued with Whewell in 1837, that present forces work; that they are, in their accumulations, *verae causae.* A former greater intensity in geological forces was not impossible perhaps, but it was conjectural. If bias there must be, then a bias toward uniformity was "more philosophical."[46] Uniformitarianism, then, was a positive operational postulate involving (1) the assumption of uniformity in the operation of the laws of nature, and (2) an assumption of the uniformity through time of presently observable causes and the degrees of their activity. Charles Lyell, of course, did not invent this doctrine. James Hutton, William Playfair, Lamarck, and others had defended it earlier. And it was influential, not because of Lyell's specific geological applications—his insistence on a steady-state earth was soon abandoned in the face of organic evolution and continental glaciation theory—but because it gave scientists a clear and positive formulation of a causal network and a workable mode of testing hypotheses that excluded both supernatural causes and unknowable hypothetical natural causes. A "wave of translation" and a miracle were both beyond the working reach of science. Uniformitarianism encouraged geologists to abandon such theories.

An intriguing question that emerges from the operational nature of Lyell's geology is the extent to which he believed that the steady-state model was a true picture of the earth's past, as opposed to being only a useful approach to solving geological questions. The ability to work with ideas that might be only approximations of the truth or not true at all was one of the most distinctive features of the new episteme and the one most baffling to its critics. Several historians have suggested that Lyell's doctrine was intended to open up geological research and not to provide a cosmology.[47] There is, in fact, some evidence that his own beliefs about the history of the earth were quite contrary to the implications of his geological theory. He distrusted speculations about "beginnings" not because he denied the existence of a beginning, but because they inhibited or misdirected geological research. He protested that his was not a doctrine of eternal recurrence or of an eternal earth. He accepted a finite earth as probable on analogical grounds—"a metaphysical question, worthy of a theologian," he said—but insisted that such ideas were useless in science and that, consequently, science could shed no light on such problems. He did not deny that there might be "some general laws" governing "the creation of new worlds" that were different from those men experienced in their own world, but nothing could be known of them. The supposition that the nature of such "beginnings" could be inferred analogically from the present was "purely gratuitous."[48] If the steady-state changeless earth was not real for Lyell, but rather a methodological postulate, then Michael Bartholomew's argument—that it was the prospect of the descent of man from a lower form that turned Lyell away from transmutationism and its corollary of a progressive history of life on the earth, and not the logic of his steady-state theory—would be even more convincing. If true, this would also make explicable Lyell's curious relationship to religious orthodoxy, and those theological features of the *Principles* so strange that they are usually ignored. If Lyell's geology was simply a way to do geological work and not a history of the earth, then it would be clear how he could accommodate such deviations from the assumed natural order as the special creation of man and a "preternatural" flood. These unique events did not affect the continuity of geological processes and had no proper analogues within geology. Until such time as evidence could be produced to justify a belief in analogous discontinuities in the natural order, the geologist was unjustified in assuming them. Lyell's geology was a closed system of assumed physical causes logically related and analyzed. Like Baden Powell, he believed that, whatever merits this system might have in science, it could not resolve questions that were properly outside its sphere.[49]

Lyell, like so many others, was a creationist struggling to come to terms with an increasingly dominant positivism within his science. While enunciating the principles of a positive geology, he was, personally, reluctant to give up the assurance of the creationist episteme and to surrender its inhibitions. The limitations of Lyell's uniformitarianism, its inability to discover beginnings and ends, had a metaphysical dimension allied to the thought of Sedgwick and Agassiz: an unwillingness to probe the ultimate secrets of nature. Darwin's first attempt to formulate his theory echoed Lyell's caution: "Extent of my theory—having nothing to do with first origin of life, grow(th), multiplication, mind & (or with any attempt to find out whether descended from *one* form & what the form was [*sic*]." Nor was he, any more than Lyell, to be successful in escaping completely the legacy of a creationist episteme.[50]

Faced with an anomalous model of actualism in geology, Darwin for the most part followed Lyell. His own geological work had closely adhered to that pattern. But his biology was from the first developmental. While drawing on Lyell's "absolute actualism" for methodological guidance, he never entertained a steady-state view of the history of life. The prescription laid down by Sedgwick could have been written with Darwin in mind:

> We have the living world before us, and from it we derive all our fundamental knowledge of the organic laws that govern the living world. Starting with this knowledge, we pass onwards, and draw forth more extended truths by irresistible analogy, and by the application of principles well-established in the natural world.

Darwin's basic actualistic premise was that species originated by natural birth: this was the only known way in which organic beings came into existence. Some "deep organic bond," he wrote further, united allied species in time and space. "This bond is simply inheritance, that cause which alone, as far as we positively know, produces organisms quite like each other, or, as we see, in the case of varieties, nearly alike." He stressed slow and gradual change more than Lyell, who had devoted the first volume of the *Principles* to showing what sudden geological changes could be introduced by cataclysmic earth forces, but he was not opposed to saltative development as such on a priori grounds. Sudden leaps, appearances, and extinctions, though very rare, might happen. He willingly accepted any degree of saltus "which we meet with under nature, or even under domestication." The disappearance of species or whole groups of beings might be "wonderfully sudden" due to geological changes. Darwin, then, achieved a working balance between uniformitarian and catastrophist forms of actualism by acknowledging the remote possibility of a rare saltative transmutation occurring, but insisting on a *working* theory of a slow and gradual change by means of the effective causes that could be studied in the present: variation, heredity, rates of reproduction, the struggle for existence, geographical distribution, and so forth. He was impatient with demands for past histories of how this or that line or organ had developed. The theory could only be constructed from present processes and so, clearly, could only be verified by them.[51]

Darwin's application of these principles to particular scientific problems seems to have taken shape in the early period of his species work and to have changed little in later years. Surrounded by "inductionists," he was not always confident of the propriety of his practice. Thomas Kuhn has remarked that "all crises begin with the blurring of a paradigm and the consequent loosening of the rules for normal research." In the present case, those who drifted away from special creation also showed a tendency to abandon "induction" as normal scientific method. Darwin embodied the innovative use of "hypothesis" at its best, but he never fully accepted its philosophical implications, nor did he completely overcome the inhibitions of one who knew that he was innovating and necessarily violating the supposed Baconian methodological canons of his time: "I am quite conscious," he wrote to Asa Gray on the eve of the publication of the *Origin,* "that my speculations run quite beyond the bounds of true science." When Henry Fawcett, who had favorably reviewed the *Origin,* reported in 1861 that John Stuart Mill had characterized Darwin's book to him as being "in the most exact accordance with the strict principles of logic [and that] the method of investigation [was] the only one proper to such a subject," Darwin was relieved. "Until your review appeared," he added, "I began to think that perhaps I did not understand at all how to reason scientifically." He was similarly thankful to Baden Powell for a few kind words, for he suffered much at the hands of mathematicians, who usually, like so many of his critics, approached the *Origin* as if it were a proof of evolution, which of course it was not. Its supporters, on the other hand, commonly viewed it correctly as a hypothesis, based on plausibly ordered evidence and heuristic in purpose.[52]

For Darwin, then, explanatory theory was equally as important in scientific inquiry as fact-gathering, and the test of the truth of a theory was its ability to group facts under a single generalization. "I believe in the truth of the theory [of natural selection], because it collects under one point of view, and gives a rational explanation of, many apparently independent classes of facts," he wrote in 1868.[53] It seemed incredible, he told Hugh Falconer, that "a false theory would explain, as it seems to me it does explain, so many classes of facts." And yet, this was not an invitation to ad hoc theorizing. Darwin usually had a keen eye for such, even in the case of his own efforts, for instance pangenesis.[54] If ad hoc explanations, which did, after all, "explain," were not acceptable, what was? Again, following the principles of positive science, the explanation had to be within the bounds of natural causation and had to employ causes and processes known or believed on good evidence to occur. Any hypothesis that met these two criteria could be held provisionally as work went on, and then modified if necessary. Like (in his veiw) much of Lamarck's work, Darwin's own theory of pangenesis violated the second of these rules. He recognized that it was, to some extent, a *pis aller,* and his defenses of it often sound almost guilty.[55] Natural selection, he thought, met both criteria; special creation met neither. It merely verbally accounted for species; it "explained" nothing.

"Fresh creations is mere assumption," the young naturalist had written in his first transmutation notebook, "it explains nothing further; points gained if any facts are connected."[56] This insight pointed the way toward the importance of theory as a means of gaining knowledge as well as organizing that which one already had. A good hypothesis need not wait on complete knowledge. Once plausible, without waiting to be proved a *vera causa,* it could direct further research and transform all that it touched. John Tyndall described the method vividly: the investigator of nature, he wrote,

> ponders the knowledge he possesses, and tries to push it further; he guesses and checks his guess; he conjectures, and confirms or explodes his conjecture. These guesses and conjectures are by no means leaps in the dark; for the knowledge once gained casts a faint light beyond its own immediate boundaries. . . . The force of intellectual penetration . . . is not, as some seem to think, dependent upon method, but upon the genius of the investigator. There is, however, no genius so gifted as not to need control and verification. . . . Thus the vocation of the true experimentalist may be defined as the continued exercise of spiritual insight, and its incessant correction and realisation.

Descent with modification, Darwin realized early, would not only explain the origin of species, but, if pushed, would revolutionize biological science.[57] While admitting to Asa Gray that the *Origin* was "grievously hypothetical," he nonetheless told Huxley that his views would prevail among younger scientists because of their utility in "group[ing] facts and search[ing] out new lines of investigation better on the notion of descent, than on that of creation," and was greatly pleased when his "hypothetical notions" led "to pretty discoveries."[58]

For Darwin, good observing always went hand in hand with good theorizing. He explained the preponderance of geologists among his early converts on the grounds that they, unlike "simple naturalists," were "more accustomed to reasoning," and, in his short treatise on geological inquiry written for an Admiralty handbook in 1849, advised beginners to "acquire the habit of always seeking an explanation of every geological point met with: for one mental query leads to another." But deduction alone was hazardous. Hypotheses had to be continuously verified by facts. Accuracy, he told the readers of the same essay, is one of the best ways to rein in the imagination which "is apt to run riot when dealing with masses of vast dimensions and with time during almost infinity." Darwin was ever-conscious of the seductiveness of a favored theory and made a point of recording contrary instances, failed experiments, and other negative evidence. "I have," he wrote in his *Autobiography,* perhaps with some extravagance, "steadily endeavored to keep my mind free, so as to give up any hypothesis, however much beloved (and I cannot resist forming one on every subject), as soon as facts are shown to be opposed to it." He objected strongly to Herbert Spencer's method:

> His deductive manner of treating every subject is wholly opposed to my frame of mind. His conclusions never convince me: and over and over again I have said to myself after reading one of his discussions, "Here would be a fine subject for half-a-dozen years' work." His fundamental generalizations (which have been compared in importance by some persons with Newton's laws!)—which I daresay may be very valuable under a philosophical point of view, are of such a nature that they do not seem to me to be of any strictly scientific use. They partake more of the nature of definitions than of laws of nature. They do not aid one in predicting what will happen in any particular case. Anyhow they have not been of any use to me.

Sharp observation, he counseled his son Horace, was the key to discoveries in science. Theory should guide observation, but it could also distort it. Emma Darwin reported one of her husband's favorite sayings to be, "It is a fatal fault to reason whilst observing, though so necessary beforehand and so useful afterwards."[59]

Darwin suffered from the unwillingness of his inductionist critics to consider this spreading but still novel approach to research in natural history. If it be proper to "invent" the undulatory theory of light, he complained to J. S. Henslow in 1860, why could he not "invent" natural selection? A decade later he was comforted by John Tyndall's defense: "How well Tyndall puts the 'as if' manner of philosophizing, and shows that it is justifiable. Some of those confounded Frenchmen have lately been pitching into me for using this form of proof or argument." A review of the *Origin* by F. W. Hutton, in which Hutton praised natural selection as "a very probable hypothesis—more probable than any yet brought forward—and one that by the clear and comprehensive views it gives of organic life, will lead to great discoveries," received Darwin's gratitude: "He is of the very few who can see that the change of species cannot be directly proved, and that the doctrine must sink or swim according as it groups and explains phenomena. It is really curious how few judge it in this way, which is clearly the right way." And yet, despite his indignation at those who objected to his "inventing" natural selection, Darwin remained enough of a Baconian to take pride in the fact that, while he *might* have legitimately invented it, he did not: he was "led to it by studying domestic varieties." Nor was the spread of the "'as if' manner of philosophizing" without its ironies. Hooker believed that the taxonomist who was an evolutionist must ignore his theory and proceed "as if" species were immutable; while Lyell reported that some antievolutionists had turned the tables on Darwin and admitted that while his theory was not true, it might be useful, "often suggesting good experiments and observations, and aiding us to retain" facts. Some even argued, as Darwin did regarding the illusion of design in nature, that the true cause of speciation, whatever it might be, had "effects of such a character as to imitate the results which variation, guided by natural selection, would produce."[60]

The refusal to accept his way of doing science was not unrelated to the reluctance to appreciate the merits of his argument for descent with modification. The two stood or fell together. Through the two quotations which served as mottoes in the first edition of the *Origin,* one from Whewell and one from Bacon, Darwin asserted his basic beliefs about science: (1) that all nature is to be explained by unvarying law, and (2) that there is no proper limit within nature to the inquiries of science. The *Origin* was, in effect, a manifesto for positivist science. As such, the *Origin* was profoundly incompatible with special creation. Darwin, therefore, was correct to assess this viewpoint, whether held as a belief in miracle or as a belief in creation by law, to be an obstacle to progress in biology. His purpose became the destruction of the belief in special creation—or what was left of it—and to persuade those who already had all but given it up to abandon their evasive retreat into nescience and to embrace openly the doctrines of descent with modification. Darwin thought it inevitable that someone would advance a viable theory of descent. "The sooner the battle is fought," he wrote to Gray, "the sooner it will be settled."[61]

4 Special Creation in the *Origin*: the Scientific Attack

In the *Origin* Darwin explored what he considered to be the deficiencies of special creation from a perspective that combined two realms of intellectual experience: positive science and theology.

During the *Beagle* voyage and in the 1839 edition of the *Journals* (written in 1837) he had employed special creationist explanations in a conventional way.[1] Although it is customary (and not unreasonable) to identify his full conversion to evolution with his first notebooks on transmutation, one should not ignore his repeated statement that it took him some years to accept unreservedly the idea of descent with modification.[2] To experiment with various hypotheses and to be firmly committed to a belief are not the same thing; and Darwin certainly had the type of intellect that could play with ideas in this manner. By the time of the writing of the sketch of 1842, however, if not earlier, there is every indication that he had completed the process of transformation. There he argued forcefully that if species had been individually created, they would not appear to have evolved from other forms.[3] This conclusion rested on the principles of analogy and the uniformity of nature, the two pillars that made positive science possible. It also required a theological support.

In rejecting special creation Darwin sought to remove the cloud of inexplicable mystery surrounding the question of the succession of species and make possible further research into the natural processes involved. "It accords with what we know of the laws impressed by the Creator on matter," he wrote in 1844, "that the production and extinction of forms should, like the birth and death of individuals, be the result of secondary means." Neither a belief in inscrutable supernatural causes nor an attitude of nescience regarding the origin of species could be allowed to discourage inquiry.[4]

A third and related obstruction was the idealism of naturalists like Owen and Agassiz which, in Darwin's opinion, gave only an illusion of knowledge. Terms such as "plan of creation" and "unity of design" which characterized the discourse of creationism were merely facades behind which naturalists hid their "ignorance" and thought "that we give an explanation when we only restate a fact." Despite Owen's boast that to the "most devoted disciple of the school of 'positive facts' the facts are not less 'positive' than they were before, only they cease to be empirical and become intelligible" and Sedgwick's confidence that "Final Causes" had, in the hands of Cuvier, Owen, and others, "rationalized a multitude of known truths" and "been continually pregnant with new discoveries," Darwin found these accomplishments empty at exactly that point where science should be fullest: explanation. He complained in 1838 that

> the explanation of types in structure in classes—as resulting from the *will* of the deity, to create animals on certain plans,—is no explanation—*it has not the character of a physical law* & is therefore utterly useless.—it foretells nothing because we know nothing of the will of the Deity, how it acts & whether constant or inconstant like that of man.—the cause given we know not the effect.

He saw in the curious idealist taxonomy of W. S. Macleay, which ordered all animal species in declining groups of five, the threat of rational design and noted in the margin of the 1844 essay, "I discuss this because if Quinarism true, I wrong." Genetic relation, he argued in 1842, was the "simplest explanation" of unity of type within the great classes. Rudimentary organs, he said, which are perplexing on the ordinary creationist view,

> can by my theory receive simple explanation or they receive none and we must be content with some such empty metaphor, as that of De Candolle, who compares creation to a well covered table, and says abortive organs may be compared to the dishes (some should be empty) placed symmetrically![5]

Such explanations were not only empty because they had no referent in the positive universe, but redundant as well. The authors of idealist taxonomies thought that their systems revealed the plan of the Creator, but "unless it be specified whether order in time or space, or what else is meant by the plan of the Creator, it seems to me that nothing is thus added to our knowledge." It seemed much more satisfactory to simply say that Owen's ideal archetype was a real ancestor and that the so-called metamorphic changes seen in the vertebrae, in the jaws of crabs, and in the pistils of flowers, and discussed metaphorically by naturalists, were actual hereditary alterations.[6] Biological idealism with its close affinities with special creation and nescience was a dead end no less than they.

The close epistemic ties among special creation, nescience, and idealism made the natural theology favored by these naturalists a persistent target for Darwin. He called it "the enemy" and directed his book on the fertilization of orchids against it.[7] This book, which appeared in 1862, had a triple purpose: first, to show that "the contrivances by which Orchids are fertilized, are as varied and almost as perfect as any of the most beautiful adaptations in the animal kingdom"; secondly, "to show that these contrivances have for their main object the fertilization" of each flower by the pollen of another flower; and thirdly, "to show that the study of organic beings may be as interesting to an observer who is fully convinced that the structure of each is due to secondary laws, as to one who views every trifling detail of structure as the result of the direct interposition of the Creator." Looking on "the flowers of Orchids, in their strange and endless diversity of shape" as comparable "with the great vertebrate class of Fish, or still more appropriately with tropical Homopterous insects, which seem to us in our ignorance as if modelled by the wildest caprice," Darwin argued throughout that the many homologies found among orchids were more economically explained by the supposition of a common descent:

> Can we, in truth, feel satisfied by saying each Orchid was created, exactly as we now see it, on a certain "ideal type"; that the Omnipotent Creator, having fixed on one plan for the whole order, did not please to depart from this plan; that He, therefore, made the same organ to perform diverse functions—often of trifling importance compared with their proper functioning—converted other organs into mere purposeless rudiments, and arranged all as if they had to stand separate and then made them cohere? Is it not a more simple and intelligible view that all Orchids owe what they have in common to descent from some monocotyledonous plant . . . and that the now wonderfully changed structure of the flower is due to a long course of slow modification . . .?

The heuristic uses of the evolutionary view were great for it offered solutions for numerous structural and specific puzzles.[8] The idealist view, on the other hand, he thought, left the naturalist little but the pleasure of a rapt contemplation of the unfathomable and ingenious mind of the maker.

The teleology of idealism frustrated what Darwin considered to be the proper growth of biological understanding in other ways as well. Events were given spurious purposeful explanations that blunted the need to find true causal ones, as by Hugh Miller, who while admitting, in regard to the decline in the size of reptiles since the "secondary period," that, "though we cannot assign a *cause* for this general reduction of the reptile class, save simply the will of the all-wise Creator," went on to argue that the *reason* why it should have taken place seemed easily assignable, "that is, to enable the succeeding mammals to flourish."[9] For Darwin such reasoning in science was doubly flawed. It was both teleological and sterile. The assumption that nature was the result of purposeful design and that explanations of natural processes must reveal this purpose was, for him, a chief impediment to the discovery of nature's laws. In his first transmutation notebook he wrote, "Kirby all through Bridgewater errs greatly in thinking every animal born to consume this or that thing.—There is some much higher generalization in view."[10] He never lost this skepticism. "Do you really suppose," he wrote to a friend years later, "that for instance Diatomaceae were created beautiful that man, after millions of generations, should admire them through the microscope? I should attribute most of such structures to quite unknown laws of growth. . . ."[11] His aversion to teleology in nature was so strong that he even questioned, to some degree, the propriety of using telic language in describing artificial selection:

> To use such an expression as trying to make a fantail [pigeon] is, I have no doubt, in most cases, utterly incorrect. The man who first selected a pigeon with a slightly larger tail, never dreamed what the descendants of that pigeon would become through long-continued, partly unconscious and partly methodical selection.

To those like Whewell who believed that the evidence of a rational, purposeful, and benevolent design was manifest in the universe, Darwin issued what was to him the obvious challenge: if any organic structure could be shown to have been produced with the good of some other species as its sole purpose, a condition that, in his belief as well as theirs, would imply foreknowledge, that would be the end of natural selection. Naturalists had to break free of long-accustomed, even supposedly natural, modes of perception—and what could be more natural than to believe that remarkable instincts must result from "the direct interposition of the Creator"?—and reject teleological doctrines if they were to gain a knowledge of nature's laws and processes. "The eye to this day," he wrote to Asa Gray in 1860, "gives me a cold shudder, but when I think of the fine known gradations, my reason tells me I ought to conquer the cold shudder."[12]

The removal of what Darwin considered to be the cluster of errors and inhibitions surrounding special creation, idealism, nescience, and teleology would make possible the development of a new biology. "As modern geology has almost banished" catastrophism, he prophesied, "so will natural selection, if it be a true principle, banish the continued creation of new organic beings, or any great and sudden modification in their structure." Difficulties of natural history, "which on the theory of independent acts of creation are utterly obscure," would be resolved, and that mentality which seemed stubbornly incapable of appreciating the manner in which theory could illuminate facts would disappear from the science of nature.[13]

The importance of a complete shift from special creation to descent with modification, from mystery and miracle to secondary causes, was clear to Darwin. There could be no halfway house. The conceptions of the new positive biology were so philosophically interrelated that no compromise with the old views was possible. He wrote to Lyell near the first anniversary of the publication of the *Origin:*

> I grieve to see you hint at the creation "of distinct successive types, as well as of a certain number of distinct aboriginal types." Remember, if you admit this, you give up the embryological argument *(the weightiest of all to me),* and the morphological or homological argument. You cut my throat, *and* your own throat; and I believe you will live to be sorry for it.

The assumptions of the new science were such that phenomena were not explained in any meaningful sense until they were subsumed under lawful natural processes which ruled out the mystery and caprice of creation. The quest after the *vera causa* mandated its rejection. "The whole subject" of expression, Darwin wrote years after the *Origin,* "had to be viewed under a new aspect and each expression demanded a rational explanation." Naturalists everywhere observed that "natura non facit saltum," that nature is "prodigal in variety, but niggard in innovation." Darwin asked,

> Why, on the theory of Creation, should this be so? Why should all the parts and organs of many independent beings, each supposed to have been separately created for its proper place in nature, be so invariably linked together by graduated steps? Why should not Nature have taken a leap from structure to structure?

"On the theory of natural selection," on the other hand, all was clear. In subsequent editions of the *Origin,* Darwin altered this passage to stress the absence of "real novelty" in nature. Intelligence and purpose should be more creative than nature showed itself to be. Nor could the idealist reply that continuity in nature reflected a continuity of the divine mind. For Darwin, this merely repeated the fact without explaining it.[14] Descent with modification not only explained the facts, but did so in such a fashion as to guide further research in a way that its rivals could not. It helped to clarify the general laws of descent, the puzzling problems of taxonomy, embryology, geographical distribution, homologies, and rudimentary organs, and explained the fossil record of the development of life.

Among the anomalies of heredity open to solution were such as caused "the occasional appearance of stripes on the shoulders and legs" of horses. "How inexplicable on the theory of creation," but "how simply is the fact explained if we believe that these species have descended from a striped progenitor" just as the numerous varieties of domestic pigeons have descended from the rock pigeon. Turning the tables on the natural theologians who liked to force an option between chance and intelligent design, Darwin used a chance or inheritance argument to show that reversion was the only plausible explanation for certain anatomical resemblances between humans and other animals. That such common structures should appear "by chance" he found incredible. The higher variability of those structures characteristic of certain varieties within a species and the lower variability of those common to the species as a whole were explained: "on the view of species being only strongly marked and fixed varieties, we might surely expect to find them still often continuing to vary in those parts of their structure which have varied within a moderately recent period, and which have thus come to differ." Similarly, why, on the theory of special creation, should useful parts be highly variable? If one assumed community of descent, "some light" was cast on the question. In short, recognized scientific principles of analogy and economy of hypothesis argued for descent with modification. The assumption of creation, on the other hand, made these facts a source of delusion to the student of nature and hence "a mere mockery." Darwin found that he could "as well believe that fossil shells had been created within the solid rock mocking the live shells on the beach."[15]

In taxonomy "our classifications will come to be, as far as they can be so made, genealogies; and will then truly give what may be called the plan of creation." Scientifically, he wrote in *The Descent of Man,* "an indeterminate element such as an act of creation" cannot be part of a definition of a species. That species were designedly created to be sterile *inter se* to prevent crossing was a common belief but, he found, a false one. If designed for this purpose, why should effectiveness of sterility differ from species to species or within individuals of the same species? Why should some species cross easily and have sterile offspring and others cross with difficulty and have fertile ones? Why should the Creator permit hybrids at all? The true view, Darwin urged, was that the difficulty in crossing species "is simply incidental or dependent on unknown differences, chiefly in the reproductive systems of the species which are crossed." In short, interspecific sterility had no purpose, but the assumption that it did had inhibited inquiry by prejudging the question. Another curious fact of nature that Darwin found explained by descent was that species could not "be ranked in a single file, but seem rather to be clustered round points, and these round other points." Inheritance and selection made this clear as extinction and divergence operated to bring about the observed result. Similarly, that large genera should have more varieties was inexplicable on the creation theory, but was only what one would expect from a common descent if the species of the large genera had once been varieties.[16]

The descent theory and embryology went together for Darwin. He could see no convincing answer in special creation for the homologies and developmental sequences of embryos in the same class. Although he stopped short of endorsing as a demonstrated law Agassiz's concept that development of the embryo paralleled the fossil sequence of ancestral forms and recapitulated those forms, he believed that the facts of embryology above all others required evolution for their explanation, and he expected further work in this field to be done under the guidance of that theory.[17]

Species, in their geographical distribution, "though evidently all more or less connected together," as Darwin wrote in the 1844 sketch, "must by the creationist (though the geologist may explain some of the anomalies) be considered as so many ultimate facts. He can only say, that it so pleased the Creator" that species be where they are and related as they are. But such a conclusion "is absolutely opposed to every analogy drawn from the laws imposed by the Creator on inorganic matter." The operation of laws, not mere facts, should be the naturalist's ultimate concern. The origination of each species or variety by "ordinary generation" in a single geographical location and its migration elsewhere was one of the fundamental generalizations that the descent theory could supply. "He who rejects it," Darwin said, "rejects the *vera causa* . . . and calls in the agency of a miracle." And the appeal to miracle, of course, blocked further inquiry. The great variety of organisms found in similar environmental situations, the narrow and wide ranges of closely allied species, the resemblance of island dwellers to the forms on nearby continents, could be explained by the creationist only as the result of divine caprice. But the descent theory suggested a multitude of natural reasons for these relationships.[18]

Like geographical distribution, homologies were "ultimate facts and incapable of explanation" by the special creationist save by recourse to the unacceptable theories of idealism. The descent theory, however, explained their derivation from a common ancestor. Taxonomy and paleontology were thus illuminated.[19]

Under the old dispensation the existence of rudimentary organs was one of the most puzzling facts in nature. Creationists sometimes explained them by appeals to "symmetry" or to the need to "complete the scheme of nature." But for Darwin, this was "no explanation, merely a restatement of the fact." Who would accept such an explanation for the path of satellites around their planets? How much better a reason was disuse and adaptation.[20]

Darwin early noted the importance of the fossil record to the descent theory: "If species really, after catastrophes, created in showers over world," he jotted in the margin of the 1844 sketch, "my theory false." Fortunately, he found "no grounds whatever to admit such a view." Extinctions seemed to vary with circumstances and, in a balanced natural system, analogy would suggest that species originated in the same manner. The special creationists' argument against transmutation based on the sudden appearance of new species in the geological strata, Darwin argued, was merely a mistake, itself based on a misunderstanding of the gradual and erratic processes behind the formation of the strata: one error led to another. "Each formation," he wrote, ". . . does not mark a new and complete creation, but only an occasional scene, taken almost at hazard, in a slowly changing drama." Until the fossil evidence was reasonably complete, he pointed out, negative evidence meant very little. So little was yet known of the paleontology of the earth aside from parts of Europe and North America. Darwin called attention to the great periods of time that had separated the deposition of fossil-bearing formations, to the lack of uniformity in the preservation of fossils—governed as the process was by both geological and ecological factors—to the small number and narrow range of so many species and varieties, to the confusion that migration could introduce into any fossil sequence, to the relative rarity of the preservation of land species, to the optimum chance of fossilization only in periods of subsidence of the land, to the destruction of fossils by erosion during periods of elevation, and so forth. He did not fail to mention the Catch-22 involved in any test of the descent theory by fossil sequences: that without a full record (a thing impossible in the nature of the case) "transitional varieties would merely appear as so many distinct species," and would be so classified by most paleontologists. How then could descent be shown? Darwin noted a similar paradox in taxonomy. If varieties were not fully fertile when crossed, they were classed as species. "If we thus argue in a circle," he commented, "the fertility [*inter se*] of all varieties produced under nature will assuredly have to be granted." The descent theory, on the other hand, once embraced on independent grounds, actually helped to point up the imperfection of the fossil record itself, a view already advocated by Lyell, and therefore led to a better understanding of geology. Improved geology, Darwin noted, had forced antitransmutationists, "whose general views would naturally lead them to" a catastrophist geology, to give up the notion of worldwide, periodic extinctions and creations.[21]

One should not minimize the seriousness of the fossil gaps, or the importance of fossil evidence in evolution, however; Darwin fully appreciated it. But, the fact of the matter was that geologists who appealed to the fossil record as the test of the truth of transmutation merely begged the question. They assumed that if transmutation were true the fossils would prove it, that the only reason for the absence of transition fossils was that they had never existed as living beings, that the evidence of the rocks would be definitive one way or the other.[22] Agassiz, not surprisingly, took this line; but

William Hopkins, who was among Darwin's most penetrating critics, also insisted that

> the geological evidence, which he has so entirely set at nought, is precisely that by which alone any theory like his can be adequately tested, since it is that evidence which bears directly on the great, if not the only, distinctive inference which can be drawn from such theories—the linear continuity of specific characters.

Hopkins did not say how genetic relationships could be fossilized; nor is it surprising that he dismissed Darwin's arguments drawn from geographic distribution, homologies, embryology, and so forth as irrelevant. On the other hand, Fleeming Jenkin, another of Darwin's most troublesome critics, seems to have realized that the geological record was a standoff.[23] It was one of Darwin's greatest methodological accomplishments to show the misdirection of this approach. In basing his argument not on fossils, but on living organisms—on their heredity, geographical distribution, ecology, and so on—he turned the whole controversy over transmutation largely away from the obsession with fossils, where the opponents of the theory had fixed it for decades, and toward actualistic methods and assumptions in the best tradition of Lyell.

Darwin's analysis of the fossil record, then, was not an ad hoc "explaining away" of an embarrassing absence of evidence, but a revealing of how unreasonable it was to demand that evidence because of the nature of geological processes and the youth of paleontological science. There were ways in which Darwin's theory could clearly have been falsified. He named some of them. The absence of transitional fossils, however, was not one of them. This was not a weakness in the theory, but a mistake about what constituted a legitimate test of the theory. When Darwin made the concession that the absence of fossils was the most serious argument against him, he was not only acknowledging a widespread belief which had to be catered to if it were to be overturned, but he was also admitting that if his actualistic and analogical reasoning on this point was rejected, then his theory was, indeed, in deep difficulty as far as general acceptance by the scientific community went.

Interestingly, both Agassiz and Sedgwick, prior to the publication of the *Origin,* acknowledged the imperfection of the fossil record. When Darwin turned that observation to account in the defense of the descent theory, it was startling how the testimony of the rocks improved.[24]

Darwin did not limit his scientific attack on special creation to stressing the epistemological soundness and heuristic advantages of descent with modification. Because of its influence, in one form or another, in the scientific community, special creation was sufficiently important to invite assessment on scientific grounds. It was not enough to show that the descent theory was superior science; it had to be shown to be the only science. And those creationists who, paradoxically, already granted this but continued skeptical had to be shown the scientifically pointless consequence of their position. There was perhaps no more effective way to do this than to demonstrate the scientific anomalies and futilities that resulted from a belief in special creation.

Special creation was easy to ridicule from the positivist perspective. Darwin was not always successful in resisting the impulse.[25] To view its paradoxes ironically, however, was not without critical merit. Regarding the divine contrivances, in 1844, Darwin reflected,

> It would, I think, be a marvellous fact, if species have been formed by distinct acts of creation, that they should act upon each other in uniting, like races descended from a common stock. In the first place, by repeated crossing one *species* can absorb and wholly obliterate the characters of another, or of several other species, in the same manner as one race will absorb by crossing another race. Marvellous that one act of creation should absorb another or even several acts of creation![26]

In dealing with special creation Darwin concentrated on the "ordinary view" of the doctrine, which was the tradition emphasizing the Creator's rationality, economy, consistency, and sense of order, purposefulness, and beneficence; in other words, the tradition of William Paley and the Bridgewater Treatises as updated by the idealists. But he found little in nature that would bear out this interpretation save on an ad hoc basis. The essence of the approach was mystery and caprice. As science, it concealed much more than it revealed. In 1868 he complained:

> On the ordinary view of each species having been independently created, we gain no scientific explanation of any one of these facts. We can only say that it has so pleased the Creator to command that the past and present inhabitants of the world should appear in a certain order and in certain areas; that He has impressed on them the most extraordinary resemblances, and has classed them in groups subordinate to groups. But by such statements we gain no new knowledge; we do not connect together facts and laws; we explain nothing.

"Nothing," he wrote of homologies in the *Origin,* "can be more hopeless than to attempt to explain [the] similarity of pattern in members of the same class, by utility or by the doctrine of final causes. . . ." If we invoke the "ordinary view" of special creation, "we can only say that so it is;—that it has so pleased the Creator to construct all the animals and plants in each great class on a uniformly regulated plan; but this is not a scientific explanation." On the other hand, a believer in special creation might prefer to "mask his ignorance, and say with Milne Edwards that the 'law of economy' is almost as paramount in nature, as the law of 'the diversity of products'" in explaining the rarity of the appearance of new organs. But the "law of economy is only the law of descent, [by which] the canon *'Natura non facit saltum'* becomes scientifically explicable." In the same way, in the orchid book Darwin was impressed by the "endless diversity of structure—the prodigality of resources—for gaining the very same end, namely, the fertilization of one flower by the pollen of another." Selection theory provided the explanation of this fact which was otherwise attributable only to a wild and wasteful caprice in the Creator. Why, to instance another problem, should there be a relation between the affinities of mammals living on islands and the depth of the water separating them? Darwin found this inexplicable on the ordinary view of creation. On the descent theory, however, we could formulate a generalization: owing to a gradual subsidence of the ocean floor, the deeper the water, the longer the separation, and the greater would be the divergence of species.[27] But, to Darwin, the generalizing powers of special creation seemed limited to marveling at the wisdom (or the playfulness) of God.

Darwin found it possible to formulate several hypotheses on the "ordinary view of creation" that could be confirmed or rejected empirically. As shown, the absence of transitional forms in the fossil record had long been one of the strongest arguments against transmutation. In the 1861 edition of the *Origin,* Darwin enjoyed reporting that the "assertion" that "geology has yielded no linking forms . . . is entirely erroneous." What had been, in effect, the prediction by creationists that no links would be found had failed. Accurate prediction being, for the positivist, a major test of the validity of a scientific theory, Darwin tested special creationism in this way in other instances. "It might have been expected," he wrote concerning the migration of plants, "that the plants which have succeeded in becoming naturalized in any land would generally have been closely allied to the indigenes; for these are commonly looked at as specially created and adapted for their own country." But, in reality, "the case is very different." Similarly, if interspecific sterility had been intended to prevent the crossing of related species, one would expect it to work to that end with greater efficiency than one found in nature. What one discovered was not infrequent successful crossing, and hybrids were themselves sometimes fertile. On the theory that the Creator had carefully adjusted each species to its environment one might expect the blind cave animals of America and Europe, who shared virtually identical "conditions of life" in their "deep limestone caverns under a nearly similar climate," to closely resemble each other. In fact, "this is not the case, and the cave-insects of the two continents are not more closely allied than might have been anticipated from the general resemblance of the other inhabitants of North America and Europe." Darwin concluded that "it would be most difficult to give any rational explanation" for this "on the ordinary view of their independent creation."[28]

The example of the cave animals introduced a second creationist principle that Darwin tested. The "ordinary view" often claimed that species and their environments were so closely adapted to one another that only intelligent design could explain it. Darwin, on the contrary, found the world filled with anomalous matches and unused opportunities. He commented in the 1838 notebooks on Richard Owen's difficulties in matching Australian fauna with environment, and noted in a marginal remark in the 1842 sketch that

> the supposed creative spirit does not create either number or kind which are from analogy adapted to site . . . it does not keep them all permanently adapted to any country,—it works on spots or areas of creation,—it is not persistent for great periods,—it creates forms of some groups in some regions, with no physical similarity,—it creates, on islands or mountain summits, species allied to the neighboring ones, and not allied to alpine nature as shown in other mountain summits—even different on different islands of similarly constituted archipelago, not created on two points: never mammifers created on small isolated island; nor number of organisms adapted to locality: its power seems influenced or related to the range of other species wholly distinct of the same genus,—it does not equally effect [*sic*], in amount of difference, all the groups of the same class.[29]

In *Natural Selection,* he pointed to the paradox of island flora seemingly designed for its special native environment and yet flourishing elsewhere: "Must we say that such island plants were created for the prospective chance of the island becoming joined to the mainland & then the plants in question spreading?"[30] Frogs, toads, and newts, which were often absent from oceanic islands, found the environment greatly to their liking when introduced by man. "Why, on the theory of creation," Darwin commented, "they should not have been created there, it would be very difficult to explain." Why were there so few, if any, mammals on many small suitable oceanic islands? Why do we find bats but no other species? Why do the species that are found on them usually resemble those of nearby continents even when the physical environment is very different? Why are the inhabitants of similar geographical and ecological areas often so vastly different? Darwin could find no answer to these anomalies in the assumption of a rational and economically designed world. Descent with modification and migration, however, caused everything to fall into place.[31]

Equally anomalous with species that did not live where they should and ecological areas that lacked the species that they were equipped to support were those species that were ill-adapted to the mode of life which they pursued. "He who believes that each species has been independently created," Darwin wrote in *Natural Selection,* "must feel surprise, at least I remember formerly having felt great surprise, at an animal manifestly adapted for one line of life, following another & very different line." Web-footed geese that are obviously equipped for an aquatic life yet live on land; the water ouzel, which is a member of the "strictly terrestrial thrush family" yet is active under water; a hymenopterous insect that spends hours under water but differs in no way from its terrestrial allies; woodpeckers that live on treeless pampas—such examples as these were contrary to the assumptions of the "ordinary view" of creation. To say that "it has pleased the Creator to cause a being of one type to take the place of one of another type" was hardly an explanation.[32]

Darwin had first encountered these "surprises" in South America during the *Beagle* voyage. Of a bird that "although not web-footed . . . is frequently met with far out at sea," he wrote:

> This small family of birds is one of those which, from its varied relations . . . may assist in revealing the grand scheme . . . on which organized beings have been created.

Another hint was the curious tucutuco, a mole-like rodent that lived under ground and was usually blind. "The blindness," Darwin wrote in the 1839 *Journal,* "though so frequent, cannot be a very serious evil; yet it appears strange that any animal should possess an organ constantly subject to injury." In the 1845 edition he added the reflection of how useful a knowledge of this animal would have been to Lamarck as an example of a species in the process of changing its habits and its character. Animals with pointless habits of behavior; birds that laid large quantities of eggs only to have many of them rot; bees that stung only to destroy themselves thereby—all were of a piece, "imperfections & mistakes" that were inexplicable to the special creationist, but which yielded without too much difficulty to the believer in evolution.[33] The advocate of a designed creation could only escape these dilemmas by confessing to a degree of mystery in the world that destroyed the rational union of purpose and utility that underlay his argument.

Not only did special creation fail Darwin's positivist tests of prediction and empirical verification, but it also committed the methodological error of redundancy. Darwin found creationists involved in elaborate rationalizations to reconcile their theory with the facts of nature. The insistence by some naturalists that species had been created, but that varieties were produced by secondary laws, struck him as unnecessarily complicating and obfuscating. The descent theory was attractively simpler and adequate to explain the facts:

> The conclusion that there is no essential differences, only one of degree & often in the period of variation, between Species & Varieties, seems to me at least as simple an explanation of the many difficulties by which naturalists are beset, as that each species should have been, independently created with its own system of variability,—the varieties imitating the characters of other species, supposed to have also been independently created, so closely as to defy in many cases the labours of the most experienced Naturalists.[34]

The resemblances of the offspring of crossed species and of crossed varieties he found "an astonishing fact" on the theory of specific creation, but "it harmonises perfectly with the view that there is no essential distinction between species and varieties." That species of shells had brighter colors owing to varying circumstances was a more economical hypothesis than that some were created brighter and others became so from environmental causes. Taxonomic confusions were only worsened by an insistence on a different origin for species and varieties. The strange facts of hereditary "reversion," when a newly born or immature animal showed traits associated with another species, were far more satisfactorily explained by following "the striking analogy of domestic pigeons & [attributing] all the cases to one common cause, viz. community of descent." "I believe," Darwin concluded, "that the similarity of the laws in the formation of varieties, & in the so-called creation of species, indicates that varieties & species have had a like origin; & not that the one has been due to the nature of surrounding causes, & the other to the direct interposition of the Hand of God."[35]

That Darwin could spend so much time testing special creation shows that prior to and during the time in which he wrote the *Origin* it was not an argument so alien to his mind as to be void of meaning. Positivist though he was, creationism was still a live issue for him as it was for so many of his contemporaries. Consequently he not only marshalled empirical arguments to show the scientific inutility of special creation, but he offered creationists tests to overthrow his own view. The logic of the opposition which insisted on evident purpose and design in nature as opposed to "chance" was yet strong enough in his mind for him to close the incompleted *Natural Selection* with the admission that if it could be proven that a single species had appeared at two separate localities without the possibility of migration, then "the whole of this volume would be useless & we should be compelled to admit the truth of the common view of actual creation," prompting the more strategically minded Hooker to exclaim in the margin of the manuscript,

"No, No—whether or no do not say so—it is not to the purpose." Proof, or even a high probability, that interspecific sterility did function in accord with creationist claims Darwin thought would be fatal to his theory. If it could be proven that the gaps in the fossil record were real and not merely the result of extinct but unknown forms; or that large genera consistently varied and small ones did not; if beauty could be shown to exist merely to please man; or if variety existed as an end in itself as the natural theologians said; if any species could be shown to possess a character solely for the benefit of another; if it could be shown that instincts, complex organs, or the communities of social insects could not have been formed by natural selection; if at any of these points he could be proven wrong, then Darwin acknowledged that it would be the end of natural selection and of his theory of organic development.[36]

One might think that in most of these cases Darwin was safe enough. His challenges may even appear to mock his enemies. But what he was saying was no more than this: establish your rival claims, as I have attempted to establish mine, by the accepted standards of science, or, if you cannot, admit mine until they are proved wrong. Although willing to admit it when the hard facts proved as embarrassing to natural selection as to special creation, and acknowledging the transient nature of much of the *Origin,* he never doubted that if the truth of a theory was primarily to be found in its ability to link facts together, to explain facts under laws and by true causes, then the hypothesis of descent with modification, whatever the merits of natural selection, and not special creation, was the theory that accorded best with the regularities as well as with the anomalies of nature.[37]

And yet, attacks on the traditional rationale for a belief in special creation, important though they were in drawing the lines of battle between positivism and creationism, could only carry the fight so far. As it turned out, it was not hard for those who had been special creationists to turn aside the entire effort by denouncing divine intervention themselves and passing, almost without noticing the concessions to positivism involved, from nomothetic creationism into a form of providentially guided evolution. Darwin saw in this as serious a challenge to a positive biology as special creation itself, and it occupied much of his attention after 1859. It was, however, a problem that he found much more difficult to overcome. Design was still the heart of the matter; and, for Darwin, design was a theological and moral as well as a scientific issue. A consideration of his complex relation to design, therefore, is necessary before the theological doctrines in the *Origin* can be seen in their true significance.

5 Providential Evolution and the Problem of Design

In the world of Anglo-American natural history, biblicist thought patterns and a belief in providential design were closely linked. Responding to this, Darwin did not limit the failures of special creation to its deficiencies as a source of scientific explanation. Just as Lyell had thought biblicism inhibited the development of geology as an empirically based science, so Darwin believed that the creationist-idealist mode of theorizing encouraged the continued use of biblical motives in biology. Miraculous creation, of course, was the most noticeable of these. He ended the *Origin* with an extended blast against naturalists who seemed "no more startled at a miraculous act of creation than at an ordinary birth."[1] *Per saltum* evolution, albeit by natural although mysterious means, struck him as a disguised appeal to miraculous creation, and those who used it, like his arch-critic St. George Jackson Mivart, he thought entered thereby "into the realms of miracle, and . . . [left] those of science."[2] He was bothered by the use of biblical imagery, such as the common saying among naturalists that each species had descended from a single pair. While "many general facts or laws," he argued, "indicate that each species has appeared at one point or rather area of the earth's surface, each species [was] not . . . necessarily derived from a single pair, but by the very slow modification, through selection, of many individuals of another species." A single ancestral pair might be possible in rare cases—even one parent, perhaps—but, in most cases, so many ancestors were involved in originating a new species that the idea of a single pair had no scientific value. Using such language could only mislead. And yet, so tight was the grip of the creationist tradition that Darwin himself found it as difficult to avoid using biblical images and analogies or to escape their associations as he did to avoid the language of design. He spoke of "all races of men as having certainly descended from one parent," and wrote that "all mammals must have descended from a *single* parent" to Lyell in 1860. He meant parent *stock,* of course, but ambiguity surrounds this term just as it did his different uses of the term creation. "Breaking with the Bible as a source of scientific truth" was not so simple a matter as it might seem. The well-worn habits of biblicist groupthink were hard to throw off, even for

Darwin. In *The Descent of Man* he specifically blamed a residual creationism for his overstressing utility in his early discussion of natural selection.[3]

It was the strong presuppositions of creationism in the minds of naturalists, as much as anything, that explained Darwin's extraordinary difficulty in getting his colleagues to see his meaning in his use of the term natural selection. Misunderstanding was so prevalent that he would have thought, "My brains were in a haze," had not Lyell, Hooker, Gray, and Huxley grasped his meaning. Darwin acknowledged that he might be a fool, but he could not believe that his associates "are all fools together." "Those who do not understand," he concluded to Hooker, "it seems, cannot be made to understand."[4] The well-known practice of his critics of interpreting the selection metaphor literally and then accusing Darwin of invoking a conscious agent arose from the analogies of creationism. Paley had thought "design" implied a designer and Sedgwick that "law" implied a lawgiver. Similarly, "selection" must imply a selector. The deep running anthropomorphism of creationism was to some minds an insurmountable obstruction to comprehending Darwin's complex mechanism for the modification of species.

At bottom, Darwin's problem, for himself as well as his colleagues, was one of getting free of the thought patterns of creationism. Ideas of divine creation and design were nagging dilemmas never far from his mind, and his continued use of conventional biblicist language, even if employed metaphorically, helped to increase his vulnerability to doubts. When botanist William Harvey cited the instance of a mutant begonia to suggest that new species might appear suddenly and ended his article with the words, "I would consider such an origin to be as true and as miraculous a creation (not 'manufacture') of a new type as if it had pleased the Divine Creator to call up, without seed, from the dust of the ground, a new organism, by the power of his omnipotent word," Darwin seems to have been temporarily taken aback and at a loss to give an adequate answer. When his friend Hooker came forth with the reply, he joked that Hooker knew his doctrine better than he did himself.[5] But the threat posed by divinely directed evolution as a means of creation was not lost on Darwin. If established, it would have destroyed natural selection and any hope of a positive biology. And, to his chagrin, not only his hostile critics but some of his friends and followers seemed determined to establish it.

It has been generally agreed (then and since) that Darwin's doctrine of natural selection effectively demolished William Paley's classical design argument for the existence of God. By showing how blind and gradual adaptation could counterfeit the apparently purposeful design that Paley, the Bridgewater writers, and others had seen in the contrivances of nature,

Darwin deprived their argument of the analogical inference that the evident purpose to be seen in the contrivances by which means and ends were related in nature was necessarily a function of mind. This inference, in which human and divine purpose were identified, had been its whole strength as a proof. In natural selection Darwin substituted an alternative hypothesis that was both logically adequate to account for the forms of organisms and philosophically more appealing to the positive outlook.[6]

Darwin's argument that random (that is, nonpurposeful) variations could, under selective pressure, result in adaptations that were owing to neither design nor "chance" did not, of course, disprove design in any possible formulation. One could, if so inclined, continue to believe that natural processes in general were purposeful and that the ends which they accomplished were designed. Darwin downed Paley, but found himself already outflanked by the designers: design was now found in the rational order of creation and in the process, in the laws, by which a general progressive end was achieved. This view of design was not new, of course; and interestingly, it was not one that Darwin himself found entirely unacceptable. In point of fact, Agassiz and Owen had both rejected a sole reliance on Paleyesque adaptive contrivance to prove design prior to 1859. Owen showed that many vertebrates had useless bones that were not adapted to any apparent need. Agassiz also admitted that there were useless structures in nature and added that the mere adaptation of means to ends could conceivably result from the operation of blind material forces. Both, however, continued to insist on the epistemic principle that mind was the only possible explanation of the overall plan seen in nature, and that adaptations, like homologies, were best understood as a part of that plan. We have, Owen argued, no a priori grounds to expect that the limbs of vertebrates will have a homologous structure. Analogy would lead us to expect them to be as diversified as the means which men use in their manipulations of the material world. Nor could this striking continuity of form be the result of chance: "that the organic atoms have concurred fortuitously to produce such harmony" was a belief from which "every healthy mind naturally recoils." Unexpected conformity to an ideal archetype and even curious and anomalous structures argued for purpose, mind, and design, "Nothing," wrote Baden Powell, "is made in vain if it be only made to preserve unity of system. . . . the attainment of an end by apparently circuitous means for the sake of obedience to the law of unity is, in fact, the highest indication of design. . . ."[7]

But in giving up Paley's simple and direct empiricism for an idealism which substituted an intuitively perceived plan behind the adaptations for the adaptations themselves, the design argument was, to the positivist, on less relevant ground scientifically. It had, by abandoning the empirical world, removed itself from the area of science and entered that of religion. When, eventually, belief in design became an act of faith in an incomprehensible divine wisdom and its ends, the idea of design ceased to have any potential value as a scientific explanation: its assertions were unverifiable and hence useless. When the Duke of Argyll, Mivart, and other advocates of design answered the Darwinist criticism that nature showed a wasteful economy of means that was contrary to the idea of a wise and provident designer by saying that "we have no antecedent knowledge of the Creator which can possibly entitle us to form any such presumption as to His methods of operation," they were, in effect, admitting that the benign and frugal character of the Creator was not inferable from nature; rather, it was an article of faith that was held despite what nature indicated, a faith that nature was, somehow, rationally managed.[8]

It was not the *Origin* that actually destroyed the empirical form of the design argument, of course. Philosophers had done that in the realm of logic even before Paley wrote. In the realm of nature its destruction was a result of the transformation in the way in which science was practiced. The *Origin* was only a product and an exemplar of that transformation. Nor was natural selection itself vital to the decline of Paley's adaptive argument. Evolution by any undirected natural means that counterfeited purpose, whether Darwinian selection, spontaneous genetic mutation, environmental influences, or what have you, would, logically at least, have been sufficient. To the positivist, who was committed to understand all natural processes only in terms of law and natural causes, Paleyesque design was merely incredible and its idealist successor irrelevant to the tasks of science; the invocation of a divine engineer, in any capacity, was a causal redundancy.

Thus the unstable Newtonian legacy of nature as matter in motion coupled with the idea of a supervising Creator finally fell apart. Its materialist, or positive, tendencies had long been gaining ascendency and had long been an increasing source of worry to its supporters. Design was the means by which it had been anchored to a theological base. Without design, a material science was almost irresistible. The virtual disappearance of natural theology from scientific discourse by the century's end signified more than the passing of a generation of scientists who had been born and educated in a more devout era. It indicated a change in the way scientists thought about nature and science, and in the practice of science. Not impiety but positivism had banished both theological explanations and concerns from the minds of working scientists. Those scientists who still talked about design in the new century were notable largely because of their rarity, and their insistence on the point, presented mainly to lay audiences, was eloquent in bespeaking the scientific irrelevance of their message.

Nonetheless, Charles Darwin was preoccupied with design, "that endless question," throughout his life, and in that preoccupation indicated the extent to which he himself had not passed completely out of "M. LeComte's" theological stage.[9]

Darwin's relationship to the idea of intelligent design in the world was consistently ambivalent. At Cambridge, he had found Paley one of the few authors worth reading. From the earliest days of his evolutionary speculation he had recognized design as one of the major obstacles to the establishment of the descent theory. Its obscurantism outraged him: "Mayo (Philosoph. of Living) quotes Whewell as profound because he says length of days adapted to duration of sleep of man!!! whole universe so adapted!!! & not man to Planets.—instance of arrogance!!"[10] The challenges it presented overwhelmed him. "It may be objected," he wrote in the 1842 sketch in what must be the most understated difficulty in the entire body of his work, that "such perfect organs as eye and ear, could never be formed" by an unguided natural selection. After he had convinced himself that it was possible, however, he gained courage, and acknowledged in the *Origin,* that even though the development of the eye and similar structures by means of natural selection "seems, I freely confess, absurd in the highest possible degree"; yet, if the theory could account for such structures and the conditions required by the theory could be shown to exist, then, our ignorance of the actual stages of transition not withstanding, the skeptic should allow "his reason . . . to conquer his imagination" and admit that even the difficulties presented by organs of "extreme perfection" were not necessarily fatal to the theory.[11]

But despite his sustained battle against this type of teleology in scientific explanations, Darwin had not avoided its appeal in his early speculations. "When I show that islands would have no plants were it not for seeds being floated about," he wrote in the transmutation notebooks, "I must state that the mechanism by which seeds are adapted for long transportation, seems to imply knowledge of Whole World—if so doubtless part of system of great harmony." As late as 1878, he was not unsympathetic to an appeal for advice regarding subjects appropriate to a prize competition in natural theology, and acknowledged to Lord Farrer in 1881 that he was reluctant (although lacking any conviction in his reluctance) to extend to the universe itself the creative role of random chance that he saw operating without restraint or intelligent direction in the origination of species. "The birth of the species and of the individual," he wrote in *The Descent of Man,*

> are equally parts of that grand sequence of events, which our minds refuse to accept as the result of blind chance. The understanding revolts at such a conclusion, whether or not we are able to believe that every slight varia-

> tion of structure,—the union of each pair in marriage,—the dissemination of each seed,—and other such events, have all been ordained for some special purpose.[12]

Notwithstanding all of this, his distrust of design as a mode of thinking was deep.

Following the publication of the *Origin,* Darwin had a lengthy philosophical debate on the subject of design with his friend Asa Gray, who advocated a modified version of Paley's argument: the divine design of the variations worked on by natural selection. "If anything is designed," he wrote to Gray in 1861,

> certainly man must be: one's "inner consciousness" (though a false guide) tells one so; yet I cannot admit that man's rudimentary mammae, bladder drained as if he went on all four legs, and pug-nose were designed. If I was to say I believed this, I should believe it in the same incredible manner as the orthodox believe the Trinity in Unity.

And that, of course, was not the proper way to believe things in science. "You say you are in a haze," he continued,

> I am in thick mud; the orthodox would say in fetid abominable mud. I believe I am in much the same frame of mind as an old gorilla would be in if set to learn the first book of Euclid, the old gorilla would say it was no manner of use; and I am much of the same mind; yet I cannot keep out of the question.

Earlier, he had confessed the same quandary to Gray that he later told to Lord Farrer:

> I grieve to say that I cannot honestly go as far as you do about Design. I am conscious that I am in an utterly hopeless muddle. I cannot think that the world, as we see it, is the result of chance; and yet I cannot look at each separate thing as the result of Design.

At another time:

> Your question what would convince me of Design is a poser. If I saw an angel come down to teach us good, and I was convinced from others seeing him that I was not mad, I should believe in design. If I could be convinced thoroughly that life and mind was in an unknown way a function of other imponderable force, I should be convinced. If man was made of brass or iron and in no way connected with any other organism which had ever lived, I should perhaps be convinced. But this is childish writing.[13]

Darwin, who was so adept at building models to test other beliefs, found it difficult to build one to test the argument for design within a world understood in positive terms. Perhaps he was unable to imagine a convincing test because, outside of a creationist context, design made no clear sense to him. Consequently, design without a direct divine intervention such as Gray proposed, though a real challenge, was nonetheless anomalous. It would seem that this was a concept so at variance with his thought, polarized as it was on this issue between the two rival epistemes of positivism and creationism with their differing systems of science, that it could not be formulated as a manageable problem, and his arguments against it sometimes had that semifacetious or impatient tone that one reserves for opposition that baffles. This, at least, was Darwin's frame of mind when he argued with Gray in the 1860s. But such was not always the case. Design forced its way into his positive universe. The distrusted intuitions, of which he spoke to Lord Farrer, came upon him from time to time. When the Duke of Argyll told the aging scientist that it was impossible to look at the numerous purposeful contrivances in nature and not see that intelligence was their cause, Darwin "looked at [him] very hard and said, 'Well, that often comes over me with overwhelming force; but at other times,' and he shook his head vaguely, adding 'it seems to go away.'"[14] Design was the nagging doubt that never left Darwin's mind.

Providentially designed evolution, which combined the purposeful manipulation of the laws of nature envisioned by the nomothetic creationist with the progressively unfolding divine plan of idealism and the aversion to direct intervention of positivism, spread with appalling (to Darwin) swiftness after 1859. Initially, perhaps, the most challenging form of providential evolution was that found among Darwin's hostile critics who rejected natural selection as an explanation of the historical succession of species and who favored another theory of evolution which, though natural in its means, manifested in its details and ends an overall divine purpose. Of these, probably the best known and most influential naturalists were Richard Owen, the Duke of Argyll, and St. George Jackson Mivart. Although providential evolution in some form remained a popular philosophical alternative to Darwinism and to the other purely positive evolution theories which dominated biology, throughout the rest of the nineteenth century and even into the twentieth in some circles, these three early proponents of the view may be taken as typical of this belief as Darwin encountered it. The story of providential evolution, which goes back into the previous century, does not begin with them, of course. For our purpose, however, one may consider the basic argument to have been put forth by Robert Chambers in 1844 in his *Vestiges of the Natural History of Creation.*

The cardinal difference between Chambers's evolutionary view of the history of life and that of his nomothetic creationist critics was the question of divine initiative in the creation of new species. For Chambers there was no question about divine creation itself. "That God created animated beings, as well as the terraqueous theatre of their being," he wrote, "is a fact so powerfully evidenced, and so universally received, that I at once take it for granted." The question was the *mode* of creation. Like his critics, Chambers rejected the notion of "immediate exertions" of "creative power" (that is, miraculous creation). Such an idea was, he thought, anthropomorphic, wasteful, and unworthy of God. But he also rejected any sort of *direct* or intervening divine superintendence of natural processes. There was, to be sure, a general providence, but God ruled nature entirely by means of uniform "natural laws which are expressions of his will." In short, Chambers made no place in his theory for that mysterious and paradoxical nudging of nature into new channels that the nomothetic creationist advocated. In his opinion, the need for such a hypothesis rested on a myopic view of nature and God's power. Drawing on Charles Babbage's famous "calculating machine" analogy, which showed the inability of a finite inspection of a numerical series to detect with certainty the law or rule governing the quantitative development of that series, Chambers argued against the commonly received "law" of heredity that said like always produces like. There could be—and he thought there were—"jumps" in the organic series from species to species and even from class to class. But uniform law governed all: there was no more a divine intervention in nature to cause a new species than there was in Babbage's numerical series to cause a change in the magnitude in the increase of number. If men thought otherwise, they had merely mistaken the nature and scope of the law. Nomothetic creationists, in believing that God had to intervene on an ad hoc basis to introduce new species (even though by secondary means), simply ignored the power of God to plan it all ahead of time and frame the laws of nature accordingly.[15] But not only was nature lawful in its operation, Chambers argued, it was also progressive. Fulfilling the original plan, the earth had gradually been shaped and modified to accommodate higher and higher forms of life eventuating (although perhaps not terminating) in man. All this had been arranged "in the counsels of Divine Wisdom, to take place . . . under necessary modifications, and [was] carried out . . . under immediate favour of the creative will or energy."[16]

In such directional evolutionary processes Chambers found "irresistible" proof of the design of the whole. The world could only be the fruit of "an act of intelligence above all else that we can conceive"; it was a culmination having "no other imaginable source" than God's providence and foresight. The regularity of the succession of life forms through time, the careful fitting of organisms before their births to the conditions of life they were to meet, the similar "plan" on which whole groups of beings were made, the curious and prophetic rudimentary organs which seemed to anticipate structures needed by yet to be created animals higher on the scale of life—such facts as these had for Chambers no other possible interpretation than design. To be sure, the monstrous births, the sudden leaps from one species to another, on which he relied to provide his "jumps," were immediately the result of the influence of environmental conditions, but that these departures were functional or even improvements on the parent species could only be explained as the result of very wise planning.[17]

It should be clear at this point that however "naturalistic" Chambers's theory of evolution was, it was not positivistic. On the contrary, it was a providential theory of a sort that had enduring popularity. In his answer to his critics, Chambers protested that his purpose in writing *Vestiges* had not been so much to present a new theory about the origin of species—it had been that, of course—as it had been to stress the uniformity of law and the creative power of God through law. He specifically denied that these laws of nature could work apart from God. The omission of "the Divine conception of all the forms of being which these natural laws were only instruments in working out and realizing" had been a major fault in Lamarck's flawed scheme. In fact, Chambers's entire argument for the general probable truth of transmutation was theological. The scientific evidence was discussed only after the "general likelihood" of evolution had been established on theological grounds. His explanations of the nature and ends of the "secondary" mechanisms of evolution were also ultimately theological. God had planned it all beforehand and had carefully orchestrated the convergence of the various necessary factors. In sum, Chambers simply adapted Paley to transmutation. His altered version of complete design ab initio was the same argument of contrivance with a temporal dimension added. He even endorsed Paley and the Bridgewater treatises in the first edition of *Vestiges* and, although this was subsequently dropped, the structure of his argument remained the same.[18]

Richard Owen's providential evolutionism had begun to form well before the appearance of the *Origin of Species,* but not until the 1860s did he publish any extensive speculations on the subject. Prior to that time, as we've seen, his public stance was that of an opponent of transmutation or, it might be more accurate to say, an opponent of all existing theories of transmutation and a proponent of a general skepticism about the possibility of unraveling the enigma of species origination. Privately, however, the question was a constant study. But his thoughts were so private that other naturalists could only suspect the direction his mind was taking. The crude and untenable schemes of transmutation that had been put forth during and since the previous century, Owen felt, had created a great prejudice against the idea of speciation by natural means and made it necessary for anyone who would advance this hypothesis to be sure of his ground.[19] Owen presented his ideas on evolution in the concluding portion of his massive treatise, *On the Anatomy of Vertebrates* (3 vols., 1866–68). Any association with the author of *Vestiges* probably would have horrified him, but in its providential aspects his formula departed little from that advocated by Chambers twenty-four years earlier.

Soon after his period of study under Baron Cuvier in Paris in the 1830s, Owen had concluded that his teacher's demonstration of a common plan in the various departments of organic nature made it obvious that any "series of species, whether plants or vertebrates, or other groups of organisms" had resulted from "the operation of a secondary cause." But, he added at once, "such cause [was] the servant of predetermining intelligent Will." After "extensive, patient, and unbiassed inductive research," he had rejected "the principle of direct or miraculous creation" as violating the scientific idea of "natural law or secondary cause" in nature. He denounced the "miracle theory of creation" as a "theological notion" and even censured Darwin for supposedly invoking a miracle to originate life. Like so many others, Owen had been swayed by the uniformitarian logic that if species died out gradually (as paleontology showed), and because of natural causes such as the struggle for existence, analogy suggested that they had come in the same way. In other words, if they did not go out as a result of sudden miraculous catastrophes, they did not so come in. So far, Owen might seem to be on his way to a positive theory of evolution. But this interpretation would be a mistake. "Derivation," as he called his theory, relied on a "creative power" which possessed a "grandeur . . . which is manifested daily, hourly, in calling into life many forms, by conversion of physical and chemical into vital modes of force, under as many diversified conditions of the requisite elements to be so combined."[20] Owen's invocation of a divine providential will and plan was not rhetoric disguising a "naturalistic" theory; it was a necessary part of his scientific explanation.

In his introduction to the *Anatomy,* Owen gave as one of his aims in writing the book that of

> apprehending the unity which underlies the diversity of animal structures: to show in these structures the evidence of a predetermining Will, producing them in reference to a final purpose; and to indicate the direction and degrees in which organisation, in subserving such Will, rises from the general to the particular.

The "preordained, definite and correlated courses" that Owen found in the organic world, past and present, revealed, to those who would see, the unifying "'idea' or truth of reason" that encompassed all. The fundamental principles of repetition and diversity that ruled all life worked toward a predetermined end. More particularly, the derivation of species was not accidental, but was purposefully guided by a divine will. In this manner of declining magnitude, Owen moved from the sublime generalizations of Mind in nature to the most practical applications: "I believe," he wrote, "the Horse to have been predestined and prepared for Man." *Equus,* he suggested, had had a grinder which was present in its ancestor *Palaeotherium* "taken away in order to add to the space for the application of the 'bit.'" Perhaps sensitive to the increasing tendency of some to question the obviousness of these prescient innovations, he confessed, "It may be weakness; but, if so, it is a glorious one, to discern, however dimly, across our finite prison-wall, evidence of the 'Divinity that shapes our ends,' abuse the means as we may." Advancing once more the "chance or design" option so popular since the seventeenth century, the choice between atheistic atomism and providence, Owen denied that the process of specific derivation could be traced to an "accidental concurrence of environing circumstances" any more than "Kosmos depends on a fortuitous concourse of atoms." The only explanation acceptable to a scientific mind was that of a predetermining will working toward the realization of a conscious purpose. And that purpose? The progressive, increasingly sophisticated and complex adaptation of form to life, of organisms to environment, until finally man appeared. Species, in their succession through the rocks, Owen believed, showed "the continuous operation of natural law, or secondary cause, [not only] successively but progressively; 'from the first embodiment of the Vertebrate idea under its old Ichthyic vestment until it became arrayed in the glorious garb of the Human form.'"[21]

Like Chambers, Owen turned to "preordained" monstrous births, by which means offspring were adapted to "higher purposes," to explain how one species derived from another. But, unlike Chambers, he would not appeal to "any external influence" as the trigger of these mutations. An "innate tendency to deviate from parental type, operating through periods of adequate duration," he asserted, was "the most probable nature, or way of operation, of the secondary law, whereby species have been derived one from the other." This tendency was not a Lamarckian impulse for improvement; still less did it respond to any selective pressure from without; nor did it operate by "slow degrees." Should both sexes appear from the same mother—and Owen thought this was far from unlikely—changes could be "sudden and considerable," and altered structures would then lead to altered habits and new modes of life. Hence, neither "appetency, impulse, ambient medium, fortuitous fitness of surrounding circumstances, or a personified 'selecting Nature'"—all the unsuccessful, if not fantastic, suggestions of materialistic transmutationists—had anything to do with "the transmutative act."

Even though Owen could out-naturalize Darwin, or what he thought was Darwin's position, by preferring the spontaneous generation of life to an initial miraculous creation of it, he was well aware of the crucial difference between his "naturalistic" theory and Darwin's positive one. "'Natural Selection,'" he pointed out, "leaves the subsequent origin and succession of species to the fortuitous concurrence of outward conditions." Derivation, on the other hand, "recognises a purpose in the defined and preordained course" of evolution.[22] Owen's idea of evolution was inseparable from the idea of design and purpose in nature. He put his considerable prestige on the line in support of this doctrine. And his position found few supporters more loyal or outspoken than his disciple the Duke of Argyll.[23]

The old episteme, with its emphasis on the union of theology and science in the study of nature, had no more devoted champion in the years after 1859 than this Scottish nobleman. Argyll's primary strategy in his effort to preserve a unity of science and religion by means of the design doctrine was to insist that purposeful teleological language (which scientists, try as they might, seemed unable to avoid) was necessary in carrying on scientific work because a complete scientific explanation must involve the idea of purpose. "It is well worthy of observation," he wrote in the 1880s, "that in exact proportion as {scientific} phrases do avoid {teleological language}, they become incompetent to describe fully the facts of science." For example, describing the various stages in the unfolding development of an embryo as "differentiations" failed utterly to encompass the goal-directed nature of those changes. The element of preparation for future use was the linkage that connected the entire process. Scientists of positivist bent, Argyll further charged, who, in an attempt to avoid religion and metaphysics, renounced teleological concepts in their work, were only substituting a metaphysics of their own: materialism. This was a bad metaphysics because it excluded a priori evident aspects of reality. The failure in comprehension was compounded by a tendency to employ verbal juggling to conceal it. Verbal traps lay in wait for the scientist who ignored "differences of work, of function, and of result" and who exaggerated similarity of "form, or motion, or chemical composition" in order to present a reductionistic view of nature. The fullness of nature was not to be found in a "sameness of material" or in an "identity of composition" or in "mere uniformity of structure." It lay in a unity of "aims" or "action"—in short, of mind. But positive science, moved by the philosophical needs of its method, relegated these elements of nature to theology and so dismissed them from "the category of scientific facts."[24]

Admittedly there was some justification for this classification, for these elements *were* theological. But, Argyll argued, they should not be excluded from science for that reason. The concept of the supranatural, of "an Agency, which, while ever present in the material and intelligible Universe, is not confined to it, but transcends it," was a necessary part of any full understanding of the world. Those who used purposeful language—as any honest scientist would, and must—and denied its reference to mind, were giving the attributes of mind to matter.[25] The metaphysics of language, then, was an issue greatly needing analysis; and the heart of that problem was metaphor and, of course, analogy.

The essence of Argyll's interpretation of language is found in two phrases: "I hold that the unconscious metaphysics of human speech are often the deepest and truest interpretations of the ultimate facts of nature" and "all metaphor is essentially founded on the perception of analogies."[26] By way of illustration of the first, Argyll pointed out how strikingly the binomial classification system of Linnaeus had been anticipated in common speech. Uninstructed by science, the folk had separated finches into hawfinches, greenfinches, chaffinches, and so forth, with the same divisions being made in popular classifications of doves, owls, and other creatures.[27] But the second was by far the more important. Teleological language in science, the significance of which some scientists tried to evade by pleading the use of metaphor, recommended itself to them in the first place only because of its evident appropriateness. Cuvier's work had been filled with it—not from choice, however, but because "it was the automatic impression made by external facts upon the receptive structure of Cuvier's mind. He could no more have abandoned it, or departed from it, than he could have abandoned the use of speech." Similarly, "Mr. Darwin does not use this language with any theological purpose or in connection with any metaphysical speculation. He uses it simply and naturally for no other reason than that he cannot help it. . . . The greatest observer that has ever lived cannot help observing [purpose] in Nature; and so his language is thoroughly anthropopsychic," that is, teleological, as a consequence. The excuse of mere metaphorical color would not wash: the analogy was genuine, the perception true.[28]

Herbert Spencer had another explanation. "The general truth," he wrote in the *Principles of Sociology,* is "that the poorer a language the more metaphorical it is, and the derivative truth [is] that being first developed to express human affairs, it carries with it certain human implications when extended to the world around. . . ."[29] Spencer spoke, in this case, of primitive language, but his remark was equally apropos of scientific language during a "primitive" stage such as evolutionary biology was experiencing in the nineteenth century. Not the least noticeable feature of the *Origin* is Darwin's struggle to express ideas for which an adequate language had not yet been invented. As biology matured its vocabulary, metaphors and "anthropopsychisms" tended to disappear. Plausible though such speculation might have been, however, it had no effect on Argyll, for his belief in purposefulness in nature rested on foundations beyond the reach of sociology or any other science.

Argyll's belief in purpose in nature was one of the great constants in his life. From childhood to old age he never shifted or faltered in this faith. Although he employed traditional evidential arguments to demonstrate the existence of an operative will in nature, he did not hazard his own belief on the validity of such proofs, nor did he show the interest of some of his contemporaries in psychical phenomena as evidence of an independent spiritual realm.[30] Rather, for him, a knowledge of mind and intention in the structure of the world was an intuition as direct and unquestionable as his awareness of his own life. From his youth, the doctrine of mind and purpose in nature had been "not so much a doctrine as a Presence. It never appeared to me to be any mere inference or the result of an argument of any kind, however linked and strong. It was an integral part of the observed phenomena, and a direct object of perception." We immediately and intuitively recognize purpose in nature on a true analogy with our own purposes, and we instinctively know that mind is its only possible source. We know it directly and unequivocally. Argyll saw the denial of this obvious truth as the weakest sort of evasion. What men might add to this primary perception, of course, was another matter. Systems of metaphysics and theology were matters of fallible reasoning, but the possible errors of their architects did not weaken the foundation on which they built.[31]

Argyll, then, based his case for the necessity of including a teleological dimension in science on three interrelated points: the intuitive perception of the operations of mind in nature by men; the difficulty that even skeptical scientists had in avoiding acknowledging this truth in the very language they used; and the testimony of common language itself, which throughout its history showed by the unconscious use of purposive metaphors and analogies in describing the processes of nature that purpose was recognized.[32]

The very possibility of science rested on this central truth. Men can know nature, he argued, because their minds are part of it. We intuitively perceive true relationships in natural processes long before we can understand them. The technology of savages, the curious anticipations of modern science that are found in the philosophies of antiquity, and the insights of genius are so explained. "These conceptions have come to Man because he is a Being in harmony with surrounding Nature. The human Mind has opened to them as a bud opens to the sun and air." If our minds are not therein united with a greater mind, he asked, how can materialism resolve the resulting paradox?

If only man possesses mind and stands alone outside nature, how is this anomaly to be explained? Does this not violate the evident unity and harmony of nature? If man is "a child of nature" as the evolutionists say, why is he wrong when he sees in nature an "image of himself"? "The very men who tell us that we are not One with anything above us," the Duke pointed out with irony, "are the same who insist that we are One with everything beneath us. Whatever there is in us or about us which is purely animal we may see everywhere; but whatever there is in us purely intellectual and moral, we delude ourselves if we think we see it anywhere."[33] "Anthropopsychism" was the term Argyll gave to this perception of will in nature, preferring it to the less accurate (to him) term anthropomorphism, for it was the analogue of mind, not body, that men saw in nature.

> If the Universe or any part of it is ever to be really understood by us [he summed up], if anything in the nature of an explanation is ever to be reached concerning the System of things in which we live, these are the perceptive powers to which the information must be given—these [that is, purpose, will, and moral sense] are the faculties to which the explanation must be addressed. When we desire to know the nature of things "in themselves," we desire to know the highest of their relations which are conceivable to us: we desire, in the words of Bishop Butler, to know "the Author, the cause, and the end of them."[34]

The "materialist" was not, of course, left without a reply. He might protest that so far as we can tell, we are but bodies; that our "minds" are functions of our bodies; and that a "mind" separated from a body is an unclear—if not a nonsensical—notion. To which Argyll would reply: "This would be a very unsafe conclusion even if the connection between our Bodies and our Minds were of such a nature that we could not conceive the separation of the two." But such is not the case. "The universal testimony of human speech—that sure record of the deepest metaphysical truths—proves that we cannot but think of the Body and the Mind as separate—of the Mind as our proper selves, and of the Body as indeed external to it"[35]—in response to which the "materialist" might grumble that the Duke was only evading the question; for there is in the human analogy only a body/mind always united in fact even though separated, artificially, in speech, but not a disembodied mind such as Argyll sees operative in nature. Thus the disembodied mind, he might continue, is not a proper analogue of man. And to read such a mind into nature on the basis of the existence of assumed purposefulness is surely to beg the question. Adaptations of means to ends might have other causes than divine intention. The basic question here is one of material processes counterfeiting purpose.

If the "materialist" should have so argued, he would have placed his finger on a tender spot in Argyll's argument. For even though the Duke's strategy of intuition seemed to avoid the need for the elaborate but specious deductions of Paley's classical watchmaker argument, he never faced the issue raised by Darwinism as to whether a blind purposeless nature could counterfeit purposeful design. If it could, and had, then the Duke's intuitive perception of purpose could well be a delusion. And, in truth, his explanation of the universal tendency to see mind in nature was merely a clarification of the thing to be explained. To explain that men see mind in nature because they recognize therein "that special kind of agency" found within themselves is to get nowhere. The question, after all, is the validity of the perception. But because Argyll treated the perception as intuitive and irresistible he evaded the question—or mistook it. The real question raised by Darwin's work was, "Is it proper to see mind in nature as a consequence of functional adaptations?" The Duke was answering the question, "Why do men believe that mind works in nature?"[36] The one asks about adaptive processes, about means and ends, the stuff of purpose; the other is a question about human psychology.

Unhampered by such considerations, however, Argyll pressed on to turn the tables on those anthropological historians of religion who were then ridiculing his point of view (if not Argyll himself) as a primitive superstition, as a "survival" of "animism." On the contrary, said the Duke, the savage who sees the world as "full of gods" is basically right—or more so than his scientific critic, anyway. Religion was not the result of imagination or naïve psychological projection. It rested on a true perception of reality, however much it might subsequently be delusively elaborated by human fancy.[37] This placed Argyll directly athwart the main course of the scientific thought of his time. On that "deepest of all religious schisms, that which divides Animism from Materialism," as anthropologist Edward Burnett Tylor put it, he opted for the losing side.[38] Nonetheless, it was with such "animistic" doctrines that Argyll sought, as did many of his contemporaries, to salvage the design argument. He took the fate of that doctrine to be closely bound up with that of science as it had, until recently, existed: that is to say, as a work of reason.

When Argyll turned to consider nature from the viewpoint of a scientist, he saw no generalization better established than conscious contrivance. "Nothing," he assured the readers of *The Reign of Law* (1866), "is more certain than that the whole Order of Nature is one vast system of Contrivance. And what is Contrivance," he asked, "but that kind of arrangement by which the unchangeable demands of Law are met and satisfied?" Indeed, it seemed "as if all that is done in Nature as well as all that is done in art, were done *by knowing how to do it.*" That there was a unifying and purposeful plan in the history of the development of life was to Argyll as certain a fact of natural knowledge as we possessed, whatever theory one might hold to explain it. This fact, this "mental purpose and design," this "conformity to an abstract idea," was unquestionable. Men knew and recognized purpose intuitively in the adjustments and functions seen in the world of plants and animals. "The bond—the nexus—between the existence of a need and the actual meeting of that need in the supply of apparatus," Argyll wrote thirty years after the appearance of *The Reign of Law,* "can be nothing but a perceiving mind and will." Like Chambers, he extolled the power of the Creator to converge disparate causes to accomplish appointed ends. [39]

It was Darwin's greatest shortcoming that he had not seen contrivance in the process he called natural selection. It was self-evident to the Duke—who, significantly, never understood natural selection as a dynamic and historical process—that "selection means the choice of a living agent." Selection must have a selector, just as design must have a designer and law must have a lawgiver. Nature as Darwin described it seemed fortuitous, unruly, and confused. His disciples worsened things by trying to establish this principle of "chance" in an aggressive positivism that would drive design and even theism out of science. The arrangements of nature, he assured George Romanes in *Nature,* can only be explained as a result of the operation of mind. "The old school of theism is as alive as ever," he declared, and the design argument as convincing as ever.[40] Argyll was simply incredulous that anyone could be so blind. "It is impossible," he wrote, "not to see . . . that the creation of new Species has followed some plan in which mere variety has been itself an object and an aim." Consequently, Darwin's theory, with its emphasis on utility, could not be "the rule which has chiefly guided Creative Power in the origin of these new Species." A mere struggle for existence could not begin to account for the richness displayed in the organic world.[41]

Stressing purpose in nature did not, in Argyll's mind, involve any sacrifice of sound scientific principles. It was evident that "the introduction of new Species to take the place of those which have passed away, is a work which has been not only so often but so continuously repeated, that it does suggest the idea of having been brought about through the instrumentality of some natural process." The Duke did not advocate miracles. But giving up miracles did not mean giving up creation. "The adaptation and arrangement of natural forces, which can compass these modifications of animal structure, in exact proportion to the need of them," he urged, "is an adaptation and arrangement which is in the nature of Creation. It can only be due to the working of a power which is in the nature of Creative power." In abandoning the concept of a miraculous creation of each species, however, Argyll blended the supernatural into the natural world, so that all events in nature became, under law, epiphanies of the divine will and instruments of its purpose. But creation by intermediate means was still creation. It still required a divine volition for its accomplishment, even though this volition might not involve a direct intervention.[42]

Long before he accepted organic evolution (or, as he called it, development) as probably true, Argyll had rejected the idea of the miraculous intervention of God in the processes of nature, and he continued to oppose this conception of creation with as much energy as any positivist throughout his career. The creation of each new species "from inorganic matter," he wrote in the 1880s, was a notion "if not actually absurd, at least . . . very difficult of acceptance." It was "a vague conception, embodied in words which are largely metaphorical, because our very idea of an act of creation is of necessity moulded on our own performances of design and construction." At this point, Argyll seemed to come close to voiding on Spencerian grounds his own doctrine of an intuitive knowledge of purpose, but miraculous creation was not an intuitive idea. "When we come to express this conception in detail," he wrote, "we feel instinctively how difficult it is to entertain." Probably no man had ever actually so entertained it. No, intuition was on the side of secondary causes. And whatever shortcomings "development" theories might have, they at least postulated some kind of means. "I do not know on what authority it is," he commented, "that we so often speak as if Creation were not Creation unless it work from nothing as its material, and by nothing as its means." It had been the greatest achievement of *Vestiges,* he thought, to have shown that miraculous creation was "not a solution of the problem, but an abandonment of all attempts to solve it."[43]

Argyll was never satisfied with the evidence for organic evolution, and remained to the last skeptical of the theories advanced by Darwin and others to explain it. Early in his long career of opposition, he had insisted on the weakness of these theories. They lacked empirical evidence and attributed alleged effects to factors not known to operate now, or to have operated at any time in the past. There was no evidence, he charged, that new species did arise by normal birth or that environmental influences could permanently alter specific characters. Should such evidence appear, however, he saw no reason why it should not be accepted. In the meantime, the law that governed species succession seemed likely to remain a mystery. And yet, he could go part way with the evolutionists. The idea of "creation by birth" did have some explanatory merit. The presence of rudimentary organs, for example, could be plausibly accounted for in that way, as could the large number of closely allied species such as one found among wild doves. But what of such ideas as the descent of "the Whale, and the Antelope, and the Monkey" from a common ancestor? The theory lost its scientific credibility "precisely in proportion to the unlikeness of the animals to which it is applied." One need not even raise the question of man, even though homologous structures in men and apes were admittedly one of the "profoundest mysteries of Nature." Other things in nature—bilateral symmetry in all organisms from man down the scale to the Radiata, for example—could not be explained by common descent, "even in imagination." Although this severe judgment mellowed somewhat over the years, near the end of his life Argyll still believed that the "law of development" eluded human inquiry.[44]

The fundamental difficulty, of course, was the "materialistic" science practiced by most advocates of descent. Physical processes simply could not account for phenomena that were properly part of the world of mind. No amount of ingenuity could remedy this elemental misperception. But what could genuine science reveal? Not much. In *The Reign of Law,* Argyll wrote that the only certain "glimpse of Creation by Law" that naturalists could offer was two fold: first, "that the close physical connection between different Specific Forms is probably due to the operation of some Force or Forces common to them all"; and, second, "that these forces have been employed and worked, with others equally unknown, for the attainment of such ends as the multiplication of Life, in Forms fitted for new spheres of enjoyment, and for the display of new kinds of beauty." This "glimpse" was completely compatible with nomothetic creationism, and, in truth, Argyll's position on evolution in *The Reign of Law* was sufficiently ambiguous to have deceived more than one historian who mistook him for an evolutionist. But it seems clear that he did not accept evolution at that time. His position in the 1860s was one of nescience, a rejection of special creation in either form, and a slight leaning toward monstrous births as a not impossible means by which new species *might* be created but one, unfortunately, unsupported by any evidence.[45]

Thirty years later, Argyll had moved somewhat closer to the idea of evolution, but was even more deceptive to the casual reader. Reviewing Herbert Spencer's evidences for evolution drawn from the geological succession of species, classification, geographical distribution, embryology, and rudimentary organs, he wrote, "I accept all these five lines of evidence as each and all confirmatory of the leading idea of development—an idea which I hold to be indisputably applicable to everything, and especially to organic life." But this says nothing of *means*—and what did Argyll mean by "development"? He defined the term as a belief in "some things always coming to be, and other things always ceasing to be—in endless sequences of cause and effect." This certainly was not organic evolution: any nomothetic creationist would have accepted it as a fact. Argyll went on to claim that if the first forms of life had been created through the instrumentality of some unknown means, there was no reason why this act could not have been repeated during the course of the ages. In fact, he suggested that the great phyla and even the subordinate classes within them may have descended from some such ad hoc creation taking place at predetermined points in the course of the progressive development of life. These creations were not miracles, but on the question of what they were Argyll resisted committing himself. One thing, however, was clear: the Darwinists and their fellow positivists would never know the answer. "We believe," he wrote in the same essay on Spencer,

> in [a designed] organic evolution. . . . we do not believe . . . in creation without a method—in creation without a process. We accept the general idea of development as completely as Mr. Spencer does. We accept, too, the facts of organic evolution, so far as they have yet been very imperfectly discovered. Only, we insist upon it, that the whole phenomena are inexplicable except in the light of mind—that provision of the future, and elaborate plans of structure for the fulfilment [*sic*] of ultimate purposes in that future, govern the whole of those phenomena from the first to the last.[46]

It seems, then, that Argyll finally came to accept organic evolution as the best scientific hypothesis about the origin of species, but also that he never threw off the reservations of his nescient period or the philosophical presuppositions of nomothetic creationism. But if there had been evolution, there was no doubt in his mind that it had been providential evolution. "I cannot doubt," he wrote in *The Unity of Nature,* "that if Species have been begun and established through birth and ordinary generation, the rise and establishment of every variety has followed a predetermined course, and the mould of every new Organ and every new development has been implicit in every Germ." The orderly progress of life through the ages was one of the grand generalizations of biology and one of the clearest evidences of mind. For Argyll, as for other providential evolutionists, "orderly" could only mean "ordered." Also, like so many of them, he pointed to the fortunate development of *Equus* as proof of the Orderer's providential care for man.[47]

What was Argyll willing to consider as the possible means of evolution? His suggestion that pivotal innovations such as the vertebrates or mammals were introduced as germs already has been mentioned. These preformationist germs, he speculated, contained the predesigned future developments of their kind. Differentiation into orders, families, genera, and species was an unfolding of a predetermined plan. This made any sort of selection or environmental influence unnecessary and, furthermore, showed that rudimentary organs—that traditional puzzle of biology—were as likely to be organs on their way to future use as degenerations from past utility. Such "prophetic germs," as he called these incipient organs, were fatal to the "element of fortuity, which is inseparable from the idea of 'natural selection.'" Mental purpose alone could resolve these peculiar facts of nature: "they may be read either in the light of History, or in the light of Prophecy." One sees in such organs a complete foreknowledge of future forms of life and of their needs in future habitats.[48]

In *The Reign of Law,* and even while skeptical of the descent theory, Argyll had considered "new births" a means whereby one species might arise out of another. This, he thought, was a "process . . . natural and easy of conception" compared to miracles. But he acknowledged that the idea was unsupported by evidence or any known occurrence; the most one could say is that it was not impossible. Like Owen, he thought that both sexes must appear, both preadapted for whatever new purpose they were to fulfill. How such births might happen he did not know: the cause, and the end, of such innovations would have to be "extraordinary." But, for the idea to be reasonably entertained, such a means of species succession could not be separated from a directing mind. As his receptivity toward evolution increased, speciation by "leaps" became more attractive. He saw a number of theoretical advantages in the idea. Saltations would explain the mysterious gaps in the fossil record while conforming to the widespread belief that ordinary generation should, if possible, be retained as the only known mode whereby new beings came into the world. It would reconcile evolution with the great stability of species, a fact so impressive that it diminished his resistance to the idea. Like Chambers, he drew on the analogy of Babbage's calculating machine and its "law" of sudden leaps in magnitude of numeral increase.[49]

In this way, the Duke came to receive the doctrine of "creation by birth." But his hostility to Darwinism and all "materialistic" theories of evolution was not thereby mitigated. Like other providential evolutionists, he saw basic and irreconcilable differences between his perception of the world and that of positive science. With some insight, he insisted that Darwinists preferred their doctrine of "fortuity" *because* it enabled them to exclude "mind or conscious direction" from nature. "It is a cardinal dogma of the mechanical school," he wrote, "that in Nature there is no mental agency except our own; or that, if there be, it is to us as nothing, and any reference to it must be banished from what they define as science." Science on these terms was, to him, simply impossible. Physical causes alone could not account for nature as we found it. Homologies, for example, involved two questions: a "how," that is, the physical causes, and "why," the purpose served. Darwin and his followers, he charged, merged these two very different queries into one. In a sense this was true, for Darwin pursued the "how" as a historical as well as a physical question. And when one fully understood "how," say, the giraffe came to have its long neck, one also understood "why"—or "why" in the only sense in which a positivist could ask the question. For Argyll, this approach was not only incomprehensible, but it left a large part of nature unexplored and not taken into account. "Not only can we detect Purpose in natural phenomena," he argued, but "it is very often the only thing about them which is intelligible to us." Physical causes are often obscure. The pursuit of them as adequate causes might even be mistaken. In the case of birds' nests (a subject about which Argyll became embroiled with Wallace), he wrote:

> Why should a Thrush always line her nest with mud, and the Blackbird always with fibrous roots? No answer can be given. . . . I am more and more convinced that variety, mere variety, must be admitted to be an object and an aim in Nature; and that neither any reason of utility nor any physical cause can always be assigned for the variations of instincts.

Nest-building techniques, as well as coloration in birds, were to be attributed to the will of the Creator. This was the "only explanation" possible. Beauty and ornament were ends in themselves. Who could not see this? Positivists could not, and with what they felt to be good reason. One satisfied with such "mental" explanations would never be moved to find physical ones, for all his talk of secondary causes or intermediate means. But, for Argyll, the positive approach to nature resulted in theories "of purely mechanical and mindless evolution through changes infinitesimal and fortuitous." Agassiz, he pointed out, had rejected evolution rather than subscribe to such an inadequate vision of science and of nature. Providential evolution sought a third way.[50]

Argyll's antipathy to positivism and Darwinian evolution was increased by a series of bitter debates with other naturalists during the eighties and nineties. He missed hardly anyone among leading evolutionists: Spencer, Huxley, Romanes, Dyer, Lankester—the list went on. The theme was always the same: the errors of Darwinism, the folly of ignoring purpose in natural processes, and, toward the end, attacks on the personal integrity of other scientists.[51] The result was a sharp decline in the Duke's reputation as a serious thinker. He had always been treated rather lightly by the Darwin circle, but he ended his days looked on as little more than a fumbling amateur by those whom he had once considered his colleagues. W. J. Youmans summed up for the positivists in the *Popular Science Monthly.* The Duke of Argyll, he wrote, "believes in evolution, or, as he prefers to call it, development; but he wants to have it in a shape to suit himself, with little touches of special creation thrown in here and there, to ease off the difficult places and keep in touch with older modes of thought."[52] His words could stand as a characterization of providential evolution itself.

The providential evolutionist who received the greatest attention from Darwin was St. George Jackson Mivart. Mivart, who was himself a renegade Darwinian of sorts, published his *Genesis of Species* in 1871. The book enraged Darwin because of its unfair representation of his ideas on evolution, and prompted him to devote considerable space in the sixth edition of the *Origin* to a reply.[53] The volume is of interest here, however, because it continued the providential themes initiated by *Vestiges* and given contemporary prominence by Owen and Argyll.

For Mivart, the order and harmony of nature were the central indications that God's providence was manifested in the physical world. Similarly, purposefulness in organic nature, as seen in the progressive succession of life and in the adaptations of living beings, and however obscure might be its ultimate ends and unknown its means, could only be explained as the result of divine design. Mivart's intuitive awareness of design was so strong that, like Asa Gray, he found it hard to believe that Darwin really meant what he said when he denied design. And yet, separated by a generation from the Bridgewater treatises, he had to admit that design could not be proven from the data of nature. One could not find in nature such evidence of design "that no one could sanely deny it." Religious faith was the seat of the belief in design and it was in religion that science must seek an ultimate understanding of nature. Mivart took the mystery of God's relation to the world so far as to suggest that even the language of design was in error in attributing to God such human motives as were implied in the words design and purpose. When Darwin called the contradictory, or at least paradoxical, belief that the Creator had supposedly given species the ability to hybridize and then stopped it with interspecific sterility "a strange arrangement," Mivart concurred. "But this only amounts to saying," he replied, "that [Darwin] himself would not have so acted had he been the Creator. A 'strange arrangement' must be admitted anyhow, and all who acknowledge teleology at all, must admit that the strange arrangement was designed." Design, then, ended in acknowledged mystery; but that mystery, as illuminated by theology, gave structure to Mivart's science.[54]

The laws of physical nature by which means scientists described and understood the order and harmony of the world, and the laws of living nature which governed the development of both individuals and species, for Mivart, were instruments of divine contrivance through which God worked to ends known only to himself. But whatever the general end of creation might be, every means and every subordinate end (which naturalists might study) had been foreseen and planned in the original creation. Mivart seemed to sense that the need or desire to see the laws of nature as expressions of omniscient and providential will was slipping away from naturalists. Even "if deemed superfluous," he wrote defensively, the conception "can never be shown to be false."[55] Nonetheless, he, like other providential evolutionists, embraced the "naturalism" which for some, with its emphasis on law and secondary causes, had been the beginning of a draining away of belief in design, but for many others had made possible a more persuasive statement of the idea.

Like his fellow providential evolutionists, Mivart firmly rejected any sort of miraculous or, as he phrased it, "supernatural" intervention in the "whole process of physical evolution." He defined evolution as "new kinds [of species] being produced from older ones by the ordinary and constant operation of natural laws." Since he favored spontaneous generation to account for the origin of life, he was, like Owen, more "naturalistic" than Darwin himself, and there can be no doubt (notwithstanding Darwin's polemic to the contrary) that the innate "power or tendency" to which he appealed as the main cause of evolution, while vague, was not a "supernatural" influence.[56] But as in the case of the others, excluding direct and miraculous intervention did not separate nature from divine control nor free science from the responsibility to acknowledge that control.

Mivart's science was as antipositivistic as Owen's or Argyll's. Nature could not be explained or adequately described by physical causes alone. Though he rejected special creation in favor of what he called "derivative creation," Mivart remained in the creationist episteme in that he believed that true science was not possible without a theological ground intuitively perceived and imposed on physical nature by the scientist. Such doctrines of mind and transcendental purpose were those of the old science and were exactly the habit of mind against which positivists reacted.[57]

Of all Darwinism's objectionable features, that of destroying a belief in intentional creation was to Mivart the most objectionable. Darwinists, by abjuring such a creation, committed themselves to "chance," to blind "fortuity." The vision of a masterless and undesigned nature brought with it hints of ancient atomism and its attendant atheism. Mivart put great emphasis on the "accidental" aspects of natural selection and variation. He found the repeated chance convergence of biological opportunity and need simply incredible. "Is it conceivable," he asked,

> that the young of any animal was ever saved from destruction by accidentally sucking a drop of scarcely nutritious fluid from an accidentally hypertrophied cutaneous gland of its mother? And even if one was so, what chance was there of the perpetuation of such a variation?

On the other hand,

> if we accept the presence of some harmonizing law simultaneously determining the two changes [that is, innovative sucking behavior and a favorable change in the environment], or connecting the second with the first by causation, then of course, we remove the accidental character of the coincidence.

If variations had been introduced by "law," that is, in a predetermined and planned manner, Darwin's theory fell to the ground. Mivart, like Owen and Argyll, favored large variations, but he did not insist on them. Small changes, even those "insignificant and minute," were acceptable as long as they were "non-fortuitous." Darwin, Mivart noted somewhat cryptically, "does not, of course, mean to imply variations are really due to 'chance,' but to utterly indeterminate antecedents." And yet, the undetermined, even if not the result of "chance," was still not predetermined. That is to say, in Mivart's view Darwin's variations happened chaotically, undesignedly, in all directions, and totally free of divine foreknowledge and intention.[58]

Mivart's idea of evolution was very different. As he described the process, God, in establishing the laws of nature at the time of the original creation, had fixed all subsequent changes and developments. Hence the history of the development of life was an unfolding of a primeval scheme in which all the steps were predesigned to their proper ends. "New forms of animal life," he wrote,

> of all degrees of complexity appear from time to time with comparative suddenness, being evolved according to laws in part depending on surrounding conditions, in part internal—similar to the way in which crystals (and, perhaps from recent researches, the lowest forms of life) build themselves up according to the internal laws of their component substance, and in harmony and correspondence with all environing influences and conditions.

In commenting on a special phylogeny, that famous series from *Hipparion* to *Equus,* he interpreted the whole: "The series," he said,

> is an admirable example of successive modification in one special direction along one beneficial line, and the teleologist must here be allowed to consider that one motive of this modification (among probably an indefinite number of motives inconceivable to us) was the relationship in which the horse was to stand to the human inhabitants of this planet.[59]

As with other providential evolutionists, Mivart believed that saltation was the most probable means of originating new species. Differing somewhat from Owen's in that it gave a role to external conditions, Mivart's theory was that changes in the parental reproductive systems were caused by a complex of external causes and internal tendencies in those systems, which resulted in the birth of individuals significantly different from their parents. These new forms, he insisted, were not "monsters" but harmonious wholes. They could not be looked on as lucky but accidental abortions of nature. In this way, new species might appear "with comparative suddenness" as jumps involving "sensible steps, such as discriminate species from species" occurred.[60] Natural selection was not without a role in this. It destroyed "abortive and feeble attempts at the performance of the evolutionary process"; it removed the parent species from the scene; it "favours and develops useful variations" although not originating them. Sometimes Mivart wrote as if natural selection were "merely minute, indefinite variations in all directions," quite apart from any selective pressure. Technically, at least, he knew better, but the entire book was a misrepresentation of Darwin's doctrine. So stated, though, he found it impossible that the "harmonious development" of the organic world could be the result of any process dominated by natural selection. Its unruly and fortuitous operation had "to be supplemented by the action of some other natural law or laws as yet undiscovered." Saltations, on the other hand, were "orderly, and according to law, in as much as the whole cosmos is such. Such orderly evolution harmonizes with a teleology derived, not indeed from external nature directly, but from the mind of man."[61]

It was important to Mivart that a "systematic and comprehensive view of the genesis of species" should "harmonize" with philosophy and religion. He saw nothing wrong with discussing the religious implications of evolution *within* a work of science.[62] This was completely in accord with the principles of the older episteme: such a harmony of truth must exist and a true science must be predicated on its existence. But the "temple of concord" to which he sought to "add one stone" was already losing its worshippers. It was not that a form of theism consistent with positive science could not be constructed, but that there was a declining interest in doing so as a part of science itself.

Providential evolution, with its themes of law and secondary causes, with its antipathy toward miraculous intervention, was capable of plausibly counterfeiting a naturalistic or positive evolutionary science. But there were great differences between the designed evolution of Owen, Argyll, and Mivart and the views of Darwin. While Darwin acknowledged the theological origin of the laws of nature, the operation and end of those laws were not predetermined by divine will nor executed under any kind of divine supervision. It followed that there could be no point in seeking a knowledge of purpose or in giving a place to such a postulate in scientific inquiry. Providential evolution showed that design could be reconciled with descent by natural birth as the cause of new species. But from Darwin's point of view, and from the positive point of view, this was done at the cost of resting in theological explanations that were both redundant and beyond the reach of science, or, worse, by disregarding a priori any explanation based on a randomly operating mechanism of evolution. Argyll and Mivart were more sympathetic to natural selection than Owen, but all rejected its ability to produce new species. At best, they identified it with the weeding-out function given to the "struggle for existence" in pre-Darwinian natural history which preserved the purity of type, but could not create biological improvements, and viewed it with strong suspicion because of its association with the hated mindless "fortuity." The "naturalism" of these thinkers was not positivism (even though they showed its influence in giving up direct divine intervention). It was creationism, and it hid from them what Darwin found. The touchstone of positive biology, at bottom, was not law or natural causes, but the absence of conscious contrivance or purpose, however these might be conceived. A creation designed ab initio rather than ad hoc was still a creation. The appearance of a new being as a result of creative law, wrote Chambers, was "as clearly an act of the Almighty himself, as if he had fashioned it with hands."[63] For the positivists, miracle was not the only source of obscurantism. A theological understanding of admittedly natural causes, in the final reckoning, required theological explanations. Providential evolution continued the older epistemic belief that to detect divine purpose was to explain. All that was needed to turn nomothetic or nescient creationists into providential evolutionists was the acceptance of physical descent as the basis of specific origin. This was no small matter, of course, since it took them closer to positivism and out of the realm of special creation. But little else changed in their idea of science. Providential evolution, with its retention of the familiar ideas of "creation by law," the use of intermediate means to achieve predesigned ends, and its attacks on miraculous creation, had taken only a step away from nomothetic creationism and was an ultimately unsuccessful attempt to preserve the viability of a modified creationist episteme as a way of doing scientific work in a world in which evolution could no longer be denied and in which positivism could not be resisted indefinitely.

That creationism should continue in an accommodated form was not surprising, but to Darwin it was alarming when it appeared among his closest associates and supporters. If creationism and evolution could mix, could the same be said for designed evolution and natural selection?

6 Variation and the Problem of Design

The hostile versions of providential evolution which condemned natural selection or, at best, greatly reduced its effectiveness as a factor in evolution did not confront Darwin with the design idea in its most threatening form. It was among his friends, those who gave a large role to natural selection as a *vera causa* (even if less than he did himself) and who were sympathetic to the positive approach to natural history, that a more serious problem emerged. Lyell, Gray, and even, in time, Wallace proved to be as devoted to the principle of providential evolution and design as were Owen, Argyll, and Mivart. They were drawn to the question of the origin of variations which was one of Darwin's weakest points (and one not ignored by his critics). Curiously, in this they appeared more interventionist than Owen, Mivart, or even Argyll who allowed the possibility of divine interference only for his preformationist germs of each taxonomic class. These latter naturalists, with Chambers, favored a total initial creation with a subsequent unfolding of a lawful purposeful development. The providential Darwinists, on the other hand, by seizing on the variation issue as the key to design, often seemed to imply some sort of ad hoc manipulation of the laws of heredity by the Creator in the case of each new species. This continuing divine superintendence was unpleasantly analogous to the mysterious law-bound interventions of the deity of nomothetic creationism. Here, too, as in less friendly providential evolutionists, old thought patterns still worked. And they worked in ways unacceptable to Darwin for both scientific and theological reasons.

Charles Lyell, despite his vigorous assault on the popular Mosaic geology, was not opposed to design in nature. On the contrary, he insisted on it. The position he took in the first edition of the *Principles,* that the "fitness, harmony, and grandeur" of nature point to an intelligent First Cause, was never abandoned. As late as 1850, he believed that the instincts and attributes of animals later domesticated had been specially instilled in them with man's future benefit in view—which, incidentally, shows the degree to which his "naturalistic" law of creation included some sort of divine intervention. He filled his species journal with reconciliations of design and transmutation. He rejected the view of Agassiz and others that the two were incompatible and the corollary of that belief, that if secondary causes were invoked to account for new species, a directing mind must be ruled out. "Development," he wrote, is "a mode of explaining creation, not getting rid of it." He was sure that "all the species [are] foreseen, planned, not evolved by a self-acting brute machinery, for Nature is God. This may be Pantheistic," he added defensively, "but not in the sense of denying God." His belief in the constant superintendence of God in the creation caused him to reject the idea proposed by transmutationist Robert Chambers that the Creator had built various organic innovations into the laws of nature at the beginning to save time and effort—as if an eternal being "to whom Time has no existence or meaning" would need "to save trouble & intervention." He quickly centered on the source of variations as the crux of the problem: "If the Variety-making Power be not the direct presence or intervention of the Cause of Causes itself, it is the nearest to it, of any of which Man is permitted to witness the effects." Creation by designed variations, the "First Cause sustaining & creating, yet within the limits of certain laws of generation," seemed the clearest hypothesis. It recognized that truly progressive change in the series of organisms—as opposed to mere adaptation to altered conditions—"is a manifestation of power of supernatural force & wisdom & design," one "not easily reconcilable with our ordinary notions of secondary causes or of delegated powers." No, he cautioned Darwin, Paley was basically right: nature could not be explained by "deifying Matter & Force or Natural Selection." The evident presence of a divine purpose could not be ignored. It was not strange that, among those in Darwin's circle, Lyell was virtually alone in his favorable appraisal of Argyll's scientific efforts and shared his convictions about "prophetic germs" and rudimentary organs, for he, like the Duke, was still a nomothetic creationist at heart.[1]

Lyell carried his conviction of designed evolution before the public in his post-*Origin* writings. In the *Geological Evidences of the Antiquity of Man* (1863) he insisted that the "variety-making Power" was not to be identified with natural selection, endorsed Gray's pamphlet reconciling the latter with design, and even attributed his own teleological view of evolution to Darwin. In 1869, two years after he had unequivocally endorsed the descent theory in the tenth edition of the *Principles,* he wrote to Darwin that "I rather hail Wallace's suggestion that there may be a Supreme Will and Power which may not abdicate its functions of interference, but may guide the forces and laws of Nature," directing variation much as a breeder does development of his stock. In the last revision of the *Principles* in 1872, he pronounced the basic causes of evolution to be "as inscrutable to us as ever," despite Darwin's impressive efforts, and pointedly reminded his readers of the mystery of the origin of variations—a mystery involving causes "of so high and transcendent a nature that we may well despair of ever gaining more than a dim insight into them." He pointed out again that descent with modification was not a philosophy of materialism as some had mistakenly believed. Quite the contrary,

> the more the idea of a slow and insensible change from lower to higher organisms, brought about in the course of millions of generations according to a preconceived plan, has become familiar to men's minds, the more conscious they have become that the amount of power, wisdom, design, or forethought, required for such a gradual evolution of life, is as great as that which is implied by a multitude of separate, special and miraculous acts of creation.[2]

Lyell never gave up his hope of purposeful divine intervention. He merely adapted his nomothetic creationism to an evolutionary format. His last public statement found him hesitating between the old science and the new.

As already mentioned, Darwin's closest encounter with revitalized design arguments came in his exchanges with Asa Gray. Prior to his knowledge of Darwin's theory, Gray's religious orthodoxy, his easy alliance of Christianity and science, his acceptance of Paley's argument, were unshaken. He welcomed Agassiz's criticisms of Lamarck and the *Vestiges,* and, while not a biblical literalist, had little apparent aversion to biblicism in natural history. He was firmly opposed to materialistic tendencies in contemporary science. His correspondence with Darwin, in the fifties and sixties particularly, seems to have provoked a reconsideration of the design issue. He joshed Darwin about the latter's "very shocking principles and prejudices against design in nature" and expressed his determination "to baptize [the *Origin*] *nolens volens,* which will be its salvation. But if you won't have it done," he warned, "It will be damned, I fear. . . ."[3]

Aside from his concern with the *Origin,* Gray's quest for a design argument that went beyond Paley and could withstand the Darwinian logic was centered on the orchid book and Darwin's study of domestic variation. In a short review of the book on orchids, he affirmed confidently that despite its author's aversion to design, the facts of his book might be "harmonized" with many views: evolution, or direct or indirect creation. The contrivances reviewed by Darwin were, Gray claimed, "as evincive of design as are analogous arrangements in the animal kingdom, from which intention is so irresistibly inferred." Privately, he wrote to Darwin that the volume "opens up a knotty sort of question about accident or design, which one does not care to meddle with until one can feel his way further than I can." Indeed, he admitted the strength of Darwin's position on the question, but did not give up hope of "a way out." When Darwin suggested that the complex adaptations of orchid fertilization mechanisms disproved the likelihood of *per saltum* evolution—"It is impossible to imagine so many co-adaptations being formed all by a chance blow,"—Gray replied, "by any number of chance blows?" And so, he continued: "I turn the question back to you. Is not the fact that the co-adaptations are so nice next to a demonstration against their having been formed by chance blows at all, one or many?" "Here lies," he went on, "the difference between us. When you bring me up to this point, I feel the cold chill." Darwin's work gave Gray many "cold chills," but, mindful of his friend's metaphysical quandary, the botanist pressed his point: "There is design in nature, or there is not. . . . If you grant an intelligent designer anywhere in Nature, you may be confident that he had had something to do with the 'contrivances' in your orchids." "I see afar trouble enough ahead quoad design in nature," Gray acknowledged,

> but have managed to keep off the chilliness by giving the knotty questions a rather wide berth. If I rather avoid, I cannot ignore the difficulties ahead. But if I adopt your view bodily, can you promise me any less difficulties?[4]

Despite his occasional prescience, it took Gray years to begin to appreciate the real threat to design contained in natural selection. Unlike Princeton theologian Charles Hodge who was, and remains, one of the most astute writers on the theological implications of Darwin's work, Gray long hoped that Darwin could and would declare natural selection and design compatible. He had assumed, at first, that Darwin had no hostility toward design, and he himself saw no reason why he should. On the contrary, he found Darwin's case much weakened without it: "accumulated improbabilities beyond belief," he thought. Although Darwin was silent in the *Origin* on how he "harmonizes his scientific theory with his philosophy and theology," Gray assumed that Darwin was a providential evolutionist who

> regards the whole system of Nature as one which had received at its first formation the impress of the will of its Author, foreseeing the varied yet necessary laws of its action throughout the whole of its existence, *ordaining when and how each particular of the stupendous plan should be realized in effect. . . .* (italics added)

It was this last, of course, so vital to Paleyesque design, that Darwin could not accept.

Gray seemed, initially, to have been blind to the challenge of selection counterfeiting purposefulness and continued to argue the question within the boundaries which Paley had established so firmly in Anglo-American Christian apologetics. Like Paley, Gray begged the question by assuming that adaptive contrivance must be due either to "chance" or to intelligent design. If "chance" could not make the eye, then natural selection must operate under the guidance of an intelligent purpose. As a positivist, Darwin on the other hand was arguing that the eye was the product of neither "chance" nor design. Gray's inability to grasp the counterfeiting problem is curious in view of his own statement of the dilemma in 1860. In a published dialogue between two readers of Darwin (later designated "D. T." and "A. G."), Gray represented his antagonist as clearly offering this argument against design. "A. G.," however, failed to see the point. Perhaps Gray had, as he said, based the piece on a real discussion with his Harvard colleague Daniel Treadwell and so repeated Treadwell's argument without seeing its force.

By the 1870s, however, Gray had attained some insight into the counterfeit challenge, although this does not seem to have materially altered his argument. He did make attempts to stop begging the question, but continued to do so by assuming that adaptive function implied intention, whether achieved directly or indirectly. He saw that the positivist assumption of the sufficiency of natural causes to account for adaptive development also begged the question, but he failed to see that by the emerging tests of science the question was successfully begged: design was not needed to make sense of nature, and this was the only standard of judgment recognized by the new science. Gray finally gave up the attempt to demonstrate design from the data of science and fell back on faith, drawing what comfort he could from the ignorance of science, the finitude of man's understanding, and the moral superiority of the theistic hypothesis over its rival. But Gray was too committed to the traditional canons of natural theology to rely solely on fideism as a defense of God's presence in the world. Like Lyell, he saw "a way out" for design in the unknown cause of variations.[5]

Variation, Gray wrote in the 1860s, "has never yet been shown to have its cause in 'external influences,' nor to occur at random. . . . If not inexplicable, it has never been explained; all we can yet say is, that plants and animals are prone to vary, and that some conditions favor variation." He had written earlier, in his essay reconciling natural theology and natural selection,

> So long as gradatory, orderly, and adapted forms in Nature argue design, and at least while the physical cause of variation is utterly unknown and mysterious, we should advise Mr. Darwin to assume, in the philosophy of his hypothesis, that variation has been led along certain beneficial lines.

How led? Unlike Lyell, Gray did not believe that God intervened from time to time directly in the variation process. He did, however, hold that the origination of new variations was somehow under constant divine supervision. At the same time, he fully expected that variations would be shown to have physical causes. He did not see the resulting double causation as redundant, but just how a Guide guided who did not interfere with natural processes, who did not break into the causal nexus, was the question. Darwin's theory had no place for a Guide and Gray's view seemed to have no role for one except as he moved in that realm of mystery which existed at the end of every chain of scientific reasoning, the reality of which everyone had to admit. Like others, and much as he tried to avoid it by insisting that God was immediately present in all natural events, Gray saw those parts of nature which seemed to be lawless or which operated according to laws not yet known as the areas in which evident divine activity was most likely to be found. The only way that Gray could meet the positivist question, Why even postulate a designed evolution? was to show that design alone could fill some gap, for example variation, in Darwinian theory. But Gray was sufficiently a positivist himself to be unable to do this. Like many caught between the two epistemes, even though he talked as if the unknown causes of variation somehow gave God an opening, he could, when pressed, imagine only some natural sequence as the cause. In the end, the point was a conundrum no less severe than Darwin's own dilemma over divine responsibility.[6]

The weaknesses involved in Gray's attempt to wed positive science and natural theology were exposed by Darwin's disciple George Romanes who, at this time, while beginning to modify some of the positions taken earlier in his *Candid Examination of Theism* (1878), was still firmly convinced of the scientific adequacy of the new naturalistic world view. Writing in the *Contemporary Review* in 1882, Romanes said that natural selection disposed of the traditional design argument by showing that the immediate cause of organic adaptation was not divine contrivance through a direct creative act but rather a system of scientifically ascertainable causes fully sufficient to account for the phenomenon. Therefore, he charged, organic adaptations cannot be used to prove theism because we must assume theism in order to see them as a mode of creation either directly or by evolution. Nor, he added, do the adaptations of organic nature offer any more compelling evidence of design than anything else in nature. In answer to the question, posed by Gray and others, Might not evolution have been the means by which creation went on? Romanes answered:

> But what does the question amount to? It amounts to saying that if there is a mind pervading nature, evolution may have been the method in which its designs have been executed. And this is a statement which no one can dispute, so long as the question of theism is left . . . untouched. But the statement cannot be made to carry the inference that because, *on the assumption of theism,* evolution may have been (or even *must* have been) the method in which design has worked in organic nature, therefore the facts

> of organic nature furnish *evidence* of design of a quality other or better than any of the facts of inorganic nature. . . . It is one thing to show that, if we assume the existence of mind in nature, organic adaptations must be due to design; but it is quite another to prove the existence of mind in nature from the known occurrence of such adaptations.

Romanes held, further, that natural selection had no relation to schemes of natural theology or to theism:

> Whether or not there is an *ultimate* cause of a psychical kind pervading all Nature—a *causa causarum* which is the final *raison d'etre* of the cosmos—this is another question, which . . . I take to present no point of logical contact with Mr. Darwin's theory, or, I may add, with any of the methods and results of natural science.

In fact, science was damaged by the intrusion of theological doctrines into its work. "Supernatural hypotheses in general," he wrote, and "doctrines of final causes in particular" had no role to play in science. Natural theology and natural science should be kept separate and pursued separately.[7]

Romanes's article provoked Gray to an exchange of letters in *Nature* the following year. Romanes found it difficult to practice the separation of natural science and natural theology that was entailed by his "no point of logical contact" doctrine. But he fumbled his principle not in claiming improperly that natural theology was misusing scientific data, as Gray charged—for even if one grants the separation, it seems too paradoxical to deny to science the ability to point out improper uses of scientific knowledge—but by structuring a theoretical proof (which he claimed was not met) of theism and design *within* the realm of natural science. Those who radically separated science and theology, as the positivists did and as some theologians did, were, as Gray pointed out, on their own principles debarred from mixing the two. On the other hand, those who, like Gray, still saw the two as united, were put in the position of asserting (at least by implication) that theology had knowledge of the world that was denied to science. Hence, and whatever the flaws in Romanes's position (or practice), the exchange showed Gray's continuing inability really to understand the random, non-teleological nature of natural selection as Darwin conceived it and its implication for design. The main reason for this seems to have been his continuing to see theology as a meaningful part of science. Though an unwilling adherent of positivism in science, Gray did not fully accept the positivist understanding of science. He could not bring himself (despite his expectation that natural causes lay behind variations) to endorse the basic positivist assumption that when sufficient natural or physical causes were not known they must nonetheless be assumed to exist to the exclusion of other causes. Gray was, as Romanes charged, much too inclined to resort to "'the factor of intelligence' as a hypothesis whenever physical causation is found to be complex or obscure"[8]

To Darwin's vexation, even Wallace defected in the direction of design. Reviewing the tenth edition of Lyell's *Principles of Geology* in 1869, Wallace doubted that the operation of natural selection was able to account for the development of man, and followed this with a full statement of skepticism the following year. The appearance during his evolution of man's unique attributes: his consciousness, love of fine music, mathematics, and metaphysical speculation among others, Wallace argued, clearly indicated that "some other law" than natural selection, "or some other power [had] been at work." The appearance of these distinctive mental faculties *in potentia,* and especially of their accompanying physical parts, in human history long before their full employment was possible, testified to "the action of mind, foreseeing the future and preparing for it." This "superior intelligence," which had guided human evolution "for a special purpose," Wallace reflected, was not necessarily divine, nor was it necessarily outside the order of nature. Its activity certainly did not represent *miraculous* intervention in that order as miracle was commonly understood. But it was certainly something strange to the positive cosmos. Wallace was confirmed in his belief as his interest in spiritualism deepened. Indeed, spiritualism was the source of his retreat from the "materialism" of natural selection. "Preternatural intelligences," he concluded, whatever their nature, had designed human evolution and represented a hitherto unexplored realm of nature. The spiritualist hypothesis "includes and accounts for . . . the nature and variety of the facts," he wrote (echoing ironically Darwin's justification of natural selection), and was supported "by the absence of any other mode of explaining so wide a range of facts."[9]

Darwin confronted the planned variation advocates in his *Variation in Plants and Animals under Domestication* (1868). "Let," he wrote,

> an architect be compelled to build an edifice with uncut stones, fallen from a precipice. The shape of each fragment may be called accidental; yet the shape of each has been determined by the force of gravity, the nature of the rock, and the slope of the precipice,—events and circumstances, all of which depend on natural laws; but there is no relation between these laws and the purpose for which each fragment is used by the builder. In the same manner the variations of each creature are determined by fixed and immutable laws; but these bear no relation to the living structure which is slowly built up through the power of selection.

In other words, grant secondary causes as the source of variation, grant the operation of natural selection, and the imputation of purpose behind the resulting contrivance is merely redundant. Gray seemingly saw the point:

> I found your stone-house argument unanswerable in substance (for the notion of design must after all rest mostly on faith, and on accumulation of adaptations, etc.); so all I could do was to find a vulnerable spot in the shaping of it, fire my little shot, and run away in the smoke.

"Of course," he added, "I understand your argument perfectly, and feel the might of it." But, nonetheless, Gray was not willing to give up designed evolution. He welcomed the idea in others and rejoiced in any proposed modification of the Darwinian theory that would lessen the iron rule of natural selection and open up possible areas for divine activity.[10]

There were others. Geologist Joseph LeConte, erstwhile follower of Agassiz and author of the popular reconciliationist work *Evolution, and Its Relation to Religious Thought* (1888), also found the origin of variations and the early stages of the development of new organs a mystery that natural selection could not solve. Not strangely, he was also a strong believer in design:

> Design, purpose, adjustment, adaptation, are not material things, but relations or intellectual things, and therefore perceivable only as the result of thought. It is simply impossible to talk about such adaptive structures without using language which implies design. . . . It is impossible even to think of such structures without implicitly assuming intelligence as the cause.

William B. Carpenter, prominent English physiologist, in "The Argument from Design in the Organic World" (1884), endorsed the idea that variations, unexplained by Darwin, were the key to the Creator's plan and a manifestation of his indirect control over the course of evolution. In 1860 Hooker seems to have shared the view of his *Gardener's Chronicle* antagonist William Harvey that God was somehow behind variations and denied that Darwinian evolution was atheistic or that it implied a mechanistic view of nature; although, by 1866, he had suffered a sufficient change of heart to refer to his own remarks on a "wise ordinance" in nature in his British Association address as "bosh and unscientific." Nevertheless, Hooker still felt evolution a more suitable theory than direct creation for those who believed in providence. F. W. Hutton, whose later essay so pleased Darwin for its perceptive understanding of the logic of his argument, must have caused him to despair when he wrote that "we should be more inclined to refer the modifications which species of animals and plants have undergone to the direct will of God" on the ground that "a being totally ignorant of its own structure or conditions of life" would be unlikely to know how to profitably modify itself or its descendants. By the 1880s, Dana had accepted evolution, but insisted on leaving room in its processes for divine initiation. Philosopher W. Stanley Jevons advocated the view that the diversity of life forms could not be accounted for by physical laws, but required an "original act of creation," an "arbitrary choice of the Creator" at points in the course of evolution. Pragmatist F. C. S. Schiller, as late as 1897, argued that Darwin's assumption that variation was indefinite and random was a methodological assumption and, as such, was not a fact: Darwin did not *know* that variations were indefinite. There was therefore the possibility that variations were directed by an intelligent and purposeful force. Indeed, Schiller thought that the assumption that they were so directed was necessary to account for progress. Even Huxley, although denying that he subscribed to the belief himself, acknowledged that designed evolution was not inconceivable.[11]

The mystery surrounding the origin of variations was not the only source of belief in designed evolution. The widely postulated appearance of new species *per saltus,* that is, by monstrosities or "new births" or by lesser but still sizable mutations, raised the same problem of unexplained changes in, and perhaps of supernatural guidance of, the course of evolution. For Darwin, monstrous births, a doctrine favored by Chambers, Owen, Argyll, Mivart, and others, from clear theological as well as scientific motives, as an explanation of how new species, or even higher taxa, had developed, was no better than a miracle: "it leaves the case of the co-adaptation of organic beings to each other and to their physical conditions of life, untouched and unexplained." It was "no explanation" at all, of no more scientific value than creation "from the dust of the earth." Such events, if they occurred—and he thought there was no good evidence that they did in the natural state—could hardly be thought of as less than designed supernatural acts. In fact, speciation by sudden monstrosity, which necessarily presumed the perfect conformity of new species and environment, was readily conceivable only within the creationist episteme. For the creationist, species, thought of as discrete identities, could not enter the world in stages without dysfunction. The preadaptation of these supposed new species to their world was as much evidence as any reasonable mind could want of intelligent forethought and design, and their possibility was hardly credible without a belief in such design. Darwin retained enough of the old episteme in his thought to feel the force, and the threat, of this reasoning even though he rejected it. But monstrosities were only an extreme form of saltation. A lesser form, with considerable empirical validation, won adherence from even some of Darwin's supporters.[12]

The best-known comment on Darwin's much-criticized devotion to the antisaltative principle of development by small variations is Huxley's:

> Mr. Darwin's position might, we think, have been even stronger than it is if he had not embarrassed himself with the aphorism, "*Natura non facit saltum,*" which turns up so often in his pages. We believe . . . that Nature does make jumps now and then and a recognition of the fact is of no small importance in disposing of many minor objections to the doctrine of transmutation.[13]

Darwin, of course, did not deny that nature made "jumps now and then." But what was their significance? "About sudden jumps," he wrote to Harvey in 1860, "I have no objection to them—they would aid me in some cases. All I can say is, that I went into the subject and found no evidence to make me believe in jumps [as a source of new species] and a good deal pointing in the other direction." Nonetheless, "jumps" worried him. He could not see how complex structures, organs, or new varieties could arise suddenly, but the fact of strikingly large mutations could not be denied and so very little was known about them.[14] Their presence not only suggested possible inadequacy in his theory of the natural selection of slight "individual differences," but must have been disturbing on theological grounds as well. Small favorable changes were "not very wonderful" in themselves and could, under selective pressure, add up to large ones, but large favorable ones initially were hard to account for without appealing to design. Such beneficial leaps smacked too much of purpose; small random variations were more congenial to a blind natural process. The larger the jump, the greater the difficulty in explaining it. Hence, Fleeming Jenkin's celebrated (and exaggerated) impact on Darwin's thinking about "blending" inheritance, by forcing him to rely more on slight variations and less on sports as the cause of specific development, may have helped to ease the problem of accounting for large and possibly divinely initiated evolutionary leaps, just as Lord Kelvin's drastic shortening of the earth's age on physical principles raised the prospect again by increasing the need for quicker evolutionary changes, as Darwin realized and as the odious Mivart was not slow to point out.[15]

Directed or designed evolution, whether by providentially induced variations or by predetermined jumps, was unacceptable to Darwin despite its important role in winning support for the descent theory from scientists and laymen who had strong commitments to theism.[16] The restructuring of the design argument to make it adaptable to evolution and, more or less, to the practice of science along positivistic lines was an important (albeit paradoxical) step in the secularization of science and its eventual intellectual separation from theology. As it turned out, these new formulations were only "over beliefs," to use William James's expression, that played no significant part in the long-run development of evolutionary theory. They were not unimportant, however, for they eased a generation of often reluctant scientists into a "naturalistic" and ultimately positivistic world view. But, for Darwin, designed evolution, whether manifested in saltation, monstrous births, or manipulated variations, was but a disguised form of special creation. Harvey, in *The Gardener's Chronicle,* had made this clear, to his momentary discomfort. By giving God a directing role in the evolutionary process and asserting his constant supervision of that process, supporters of designed evolution introduced into science the same elements of caprice and mystery as had special creationists.

What was worse, the assumption of an active and directing deity made redundant nonsense of natural selection. To Darwin's surprise, even Gray and Wallace seemed incapable of seeing why this was so.[17] Writing to Lyell shortly after the publication of the *Origin,* Darwin firmly upheld natural selection as a positive explanation of species succession:

> I entirely reject, as in my judgement quite unnecessary, any subsequent addition "of new powers and attributes and forces"; or of any "principle of improvement," except in so far as every character which is naturally selected or preserved is in some way an advantage or improvement, otherwise it would not have been selected. If I were convinced that I required such additions to the theory of natural selection, I would reject it as rubbish . . . I would give nothing for the theory of Natural Selection, if it requires miraculous additions at any one stage of descent.

He wrote to Lyell again two years later, "The view that each variation has been providentially arranged seems to me to make Natural Selection entirely superfluous, and indeed takes the whole case of the appearance of new species out of the range of science." Against Gray's idea of guided variations, he argued that

> variations in the domestic and wild conditions are due to unknown causes, and are without purpose, and in so far accidental; and that they become purposeful only when they are selected by man for his pleasure, or by what we call Natural Selection. . . . Gray's notion of the courses of variation having been led like a stream of water by gravity seems to me to smash the whole affair. It reminds me of a Spaniard whom I told I was trying to make out how the Cordillera was formed; and he answered me that it was useless, for "God made them." It may be said that God foresaw how they would be made.

But such a belief did not give scientific knowledge; and yet, if not God, "what the devil determines each particular variation?"[18]

In admitting that the ultimate causes of variability were unknown, Darwin gave an opportunity to the advocates of designed evolution that was not lost. His use of the term chance in regard to variations was not an invocation of metaphysics but only an acknowledgment of ignorance as to what he assumed to be their lawful and natural causes. Not until near the end of his life, when he was "staggered" by the failure of a controlled manipulation of an experimental environment to induce variations even in highly variable plants, did he doubt that, in some complex way, variations were indirectly caused by the environmental "conditions of life" which, in conjunction with the internal constitution, had induced changes in the reproductive system of a parent or ancestor of the organism in question. It was this complex of causes that explained both the observed tendency to vary and the often seemingly directional nature of variations. "A similar organization being similarly acted on," Darwin theorized, led to repeated variations of the same general type. Even in the case of the baffling "spontaneous" variations which seemed quite free of this mechanism, there was no reason to abandon the assumption of a natural cause.[19] It was, however, an assumption, an extension of his positive theory. A skeptic, such as Lyell or Gray, could reasonably suggest an alternative—and did.

The origin of variations, as Howard Gruber has pointed out, was clearly the persistent question that threatened natural selection. It was also the question that threatened Darwin's positive biology. While a thoroughgoing positivist might simply have dismissed with a shrug any theories of divine intervention (and this sometimes was Darwin's impulse), Darwin's residual theism and the continuing influence of theism among his scientific colleagues made the possibility of directed evolution a live issue for him, just as special creation was a live issue. There was no real possibility, in either case, of his abandoning his commitment to a system of purely natural causes, but the challenge was no less serious for that. "An unexplained residuum of change," he uncomfortably acknowledged in *The Descent of Man,* "must be left to the assumed uniform action of those unknown agencies, which occasionally induce strongly marked and abrupt deviations of structure in our domestic productions."[20]

Lyell, Gray, and the others had to be rebutted. The strategy, however, was not clear. With no conclusive facts to support him, Darwin turned to the venerable tactic of the reductio ad absurdum. To Gray's assertion that "variation has been led along certain beneficial lines," he replied, "I cannot believe this; and I think you would have to believe, that the tail of the Fantail [pigeon] was led to vary in the number and direction of its feathers in order to gratify the caprice of a few men." He used the same example in writing to Lyell: Could he *really* think that the deity had intervened to cause variations in domestic pigeons "solely to please man's silly fancies?" But the reductio had its limits and was dependent on one's opponent sharing his assumptions. "I have asked" Lyell, he wrote to Gray, "(and he says he will hereafter reflect and answer me) whether he believes that the shape of my nose was designed. If he does, I have nothing more to say." In the same vein, he wrote to John Herschel that he could not believe that God had designed the feathers in the tail of the rock pigeon for the pleasure of the breeder, but acknowledged that "many persons" would quite freely agree that he had.[21]

Darwin, of course, was not limited to such a conditional argument. The large amount and wide range of variation in both wild and domestic species, he felt, made random selection more than plausible. To invoke lawful natural processes *and* intervention was both redundant and "mere verbiage."[22] But more was needed than a confession of faith in natural causes. He needed a theory to explain variations, and he found it in pangenesis.

"A rash and crude hypothesis," he called this theory of heredity in writing to Huxley, "yet it has been a considerable relief to my mind, and I can hang on it a good many groups of facts." To another he spoke of it as "an immense relief, as I could not endure to keep so many large classes of facts all floating loose in my mind without some thread of connection to tie them together in a tangible method." In pangenesis, Darwin assumed that each cell in the body threw off minute granules or gemmules which were then dispersed throughout the system. Having the power of self-division and of growth, the gemmules were collected in all parts of the body as well as in the sexual system. The gemmules represented the organism in all stages of growth. They were retained dormant in the body and were passed on to descendants by means of the various modes of reproduction. Cells, and so the gemmules, Darwin further assumed, were modified by "changed conditions of life." The use and disuse of organs, habit, and injury to the sexual system also modified the gemmules. By means of pangenesis Darwin felt able to account for not only heredity but such puzzling phenomena as reversion, regeneration of parts, and, most important, variability.[23]

The heuristic value of pangenesis may be fully acknowledged without believing that the atrractiveness for Darwin of what he called this "well-abused" and "universally despised" theory was thereby exhausted. I would suggest that pangenesis not only gave him a theory to hang facts on; it also filled up a troublesome hole in his positive evolutionary theory. It assured him that a capricious deity could be excluded from the processes of heredity as well as from speciation, and that Lyell, Gray, and Wallace were wrong. And even if pangenesis should fail, Darwin had prepared a second line of defense to rule out divine intervention in the process of variation. This was to catalog an impressive amount of circumstantial evidence showing the operation of regular laws. The existence of an orderly and determined natural process would then be a reasonable inference.[24]

Darwin's rejection of both direct and indirect design, special creation, and the various schemes for providential evolution meant that, unlike his associates, he had largely broken free of the prepositive view of the interrelation of science and theology. And yet traces of theological concerns remained even in his arguments against design. In his building-stone argument, in his remarks on pigeons and man's "silly fancies," and in his query about the origin of his nose, there was implicit the judgment that God would not be guilty of such frivolous conduct. Divine meddling in nature or concern with trivia at the distant creation, such as was envisioned by the advocates of designed evolution, cast God in the same capricious and even immoral role as did special creation, and the theological arguments which Darwin brought against directed evolution had been rehearsed in the *Origin.*

7 Special Creation in the *Origin:* the Theological Attack

In the *Origin of Species,* and in the writings that culminated in it, Darwin frequently condemned special creation on theological grounds. Though a positivist in science, and despite his insistence on the autonomy of science, Darwin was not able to jettison the idea of God. He needed it to underwrite the possibility of science, to guarantee its rationality, that is, the correspondence of the scientist's activity with truth, and to preserve his optimistic view of the evolutionary process. That "faith in nature as God's work, which gives to science its highest dignity"[1] and which also gave it its meaning for naturalists under the old episteme, was not entirely absent from Darwin's work. That this was inconsistent with his positive orientation in science is true but, after all, Darwin was a man, with all the contradictions of a man, and not a system. Consistency we should expect of philosophies, but perhaps not of philosophers. The *Origin* was the work of Darwin the theist as well as Darwin the positivist, and the intermingling of positivism and theology in that great work is one of its most fascinating features.

In his pathbreaking book on the origins of Darwin's theory of natural selection, Howard Gruber states:

> In Erasmus [Darwin] we find fairly frequent references to God in the deistic sense of the originator who set the world in motion according to certain general laws, but not in the sense of a designing or intervening Providence. In Charles' published writing there is really almost no mention of a Creator, even in that deistic sense.[2]

This carefully qualified statement contains a common misconception. The *Origin,* in all of its editions, not only has numerous references to such a Creator, but theological arguments based on such a conception had some importance in its overall logic as that existed for Darwin. I would be clear on this point. It is possible (and this has been, of course, the normal approach to the book) to read the *Origin* solely as a positivistic work. But to so read it is to reinterpret it from a post-Darwinian perspective. The argument of the book, in its full integrity, did not take that form for Darwin himself. During the twenty years or so in which he worked on his theory and even during the agnostic period of his later life made so familiar by his autobiography, elements of the creationist and positivistic epistemes coexisted in Darwin's mind in a loose, paradoxical, and curiously unantagonistic way. Early in his career he largely dropped theology from his science but not from his world view. As the story told by Argyll illustrates, Darwin saw the world strangely, in double vision, but in a manner that befits a man moving from one episteme to another. "I have always felt it to be a curious fact," Francis Darwin wrote of his father, "that he who had altered the face of Biological Science, and is in this respect the chief of the moderns, should have written and worked in so essentially a non-modern spirit and manner." This was truer than he realized.[3]

Darwin's theological defense of descent with modification took form in the late 1830s and early 1840s and remained consistent throughout the period of the writing of the *Origin.* It rested on four primary conceptions of God: that God cannot be the author of the cruelties and waste seen in nature; that God cannot be the Creator of a world that deceives, mocks, and misleads honest inquiry; that he has created only through general laws; and, lastly, that he does not stoop to trifling works of natural engineering.

When Charles Lyell first discussed in the *Principles of Geology* the practice of the ichneumon fly of laying its eggs in the living bodies of caterpillars that its young might feed on a living host, he showed no moral concern that nature should employ such a gruesome device. Indeed, his tone was one of implicit approval at this efficient means of checking the number of caterpillars. But by 1859 his position seems to have changed, perhaps as a result of arguing with Darwin. Then he worried about the moral ambiguity of such practices. It seemed clear that special creation or, as he called it, "direct intervention" made the Creator responsible for such beings as the Venus's-flytrap, which lured its prey to its death by perverting an instinct presumably given to the insect for its benefit. The dilemma extended to his own belief in directed variations: "It will be objected," Lyell wrote in his journal,

> that if you refer varieties to a divine source, you must attribute those which fail, monstrosities, abortions, those which meet an untimely end, to the same; useless organs, rudimentary ones, immoral, vicious individuals, evil insanity. . . .

But such puzzles were for the heads of theologians, not scientists. The latter should not, "because of this, hesitate to ascribe organic laws to the design of a Supreme Being & the order which reigns, because there are disturbances, or the good to imperfection [*sic*], because of the existence of evil."[4]

Darwin was not willing to shift the burden to the ingenuity of theologians. The moral challenge was too great. From his viewpoint, it was no real answer to the pain, terror, and waste of life in nature, to its cruelty and irrationality, to say that God had purposes that we knew not of and which were thus served.[5] Why had he such purposes? No purposes could, in Darwin's mind, redeem the *intention* to use such means. He was repelled by the natural spectacle: "What a book a devil's chaplain might write," he wrote to Hooker in 1856, "on the clumsy, wasteful, blundering, low, and horribly cruel works of nature!" And yet there was no escaping the conclusion that a designing God must be held responsible for his designs. Gray advised Darwin to silence his theological critics by taking the position that divine action in the world was "immediate, orderly, and constant." He had no conception of how repugnant to Darwin this "Christian view" was. "I cannot persuade myself," Darwin told Gray, "that a beneficent and omnipotent God would have designedly created the Ichneumonidae with the express intention of their feeding within the living bodies of Caterpillars, or that a cat should play with mice." He was indignant that the "strange and odious instinct" of the young cuckoo ejecting its foster siblings from the nest should be "boldly called a beneficent arrangement, in order that the young cuckoo may get sufficient food and that its foster-brothers may perish before they had acquired much feeling!" On the theory of natural selection, however, it was not so strange that some "contrivances of nature" that were "abhorrent to our ideas of fitness" should exist.

> We need not marvel at the sting of the bee causing the bee's own death; at drones being produced in such vast numbers for one single act, and being then slaughtered by their sterile sisters; at the astonishing waste of pollen by our fir-trees; at the instinctive hatred of the queen bee for her own fertile daughters. . . .

The moral problem was greatly lessened if blind secondary forces were the direct causes of such horrors.[6] In an often-quoted passage in the 1842 sketch, he wrote: "It accords with what we know of the law impressed on matter by the Creator, that the creation and extinctions of forms, like the birth and death of individuals, should be the effect of secondary means. . . ." This primary positivistic assumption was not merely a matter of science. Its theological significance almost immediately followed:

> We cease being astonished, however much we may deplore, that a group of animals should have been directly created to lay their eggs in bowels and flesh of other [*sic*]—that some organisms should delight in cruelty,—that animals should be led away by false instincts,—that annually there should be an incalculable waste of eggs and pollen.

Difficult as one might find it to accept such brutal intervening laws, their existence and free operation was more worthy of God than was the divine guilt that would be created by their absence.[7] The operation of these laws, "unconscious and unpitying," as he characterized natural selection, had evolved a world that was imperfect in both form and function and no better constructed than was necessary to promote the struggle for existence.[8] Lacking design and economy in its details, nature provided no evidence of that perfect rationality and wisdom argued for by creationists, and in that lack, which was to him incompatible with divine activity, Darwin attempted to find a divine exoneration as well as a support for a naturalistic theory of common descent.

A paradoxical and even contradictory feature of Darwin's moral problem in the *Origin* was his attempt to salvage something of Paley's smiling face of nature theme, notwithstanding his moral outrage and his positivistic intent to deprive nature of purpose and value. Despite his horror, Darwin, like Paley, could not bring himself to believe that nature was not ultimately benevolent, that it was not "a happy world after all."[9] His emphasis on struggle between individuals, his remorseless insistence on extinction as the fate of virtually all species, his rejection of the terms higher and lower in discussing biological organization, and his aversion to seeing evolution as "progress" are well known.[10] These were the needs of his science, but theologically he was reluctant to accept such a stark vision. In spite of his clear attempt to exonerate God from responsibility for the details of creation or to involve the creationists in a moral dilemma if His creative will and intention were insisted on, Darwin tried to look beyond the cruel world of natural selection to an end promoting the general good. "When we reflect," he wrote,

> on the struggle [for existence], we may console ourselves with the full belief, that the war of nature is not incessant, that no fear is felt, that death is generally prompt, and that the vigorous, the healthy, and the happy survive and multiply.

Similarly,

> it may not be a logical deduction, but to my imagination it is far more satisfactory to look at such instincts as the young cuckoo ejecting its foster-brother—ants making slaves—the larvae of ichneumonidae feeding within the live bodies of caterpillars—not as specially endowed or created instincts, but as small consequences of one general law, leading to the advancement of all organic beings, namely, multiply, vary, let the strongest live and the weakest die.

Of the rigors of selection pressure, he wrote in *Natural Selection,*

> we must regret that sentient beings should be exposed to so severe a struggle, but we should bear in mind that the survivors are the most vigorous & healthy & can most enjoy life: the struggle seldom recurs with full severity during each generation: in many cases it is the eggs, or very young which perish: with the old there is no fear of the coming famine & no anticipation of death.[11]

Somehow, and to whatever degree, God tempered the wind to the shorn lamb, and the ends of creation were good whatever price might be exacted by the means.

Just as Darwin did not originate the theological conundrums that appear in the *Origin,* so he did not invent the solution to the problem of evil given here. Good ends from particular evils for which God was not responsible had long been a part of Christian apologetics. It was a solution readily embraced by special creationists. But Darwin could not, any more than they, escape the paradox involved in the marriage of a lack of responsibility to omnipotence. If God did not control the means, how could the end be assured? Darwin's positivism prevented him from taking refuge in mystery, as did Sedgwick and others. The old geologist rightly accused Darwin of attempting to break the link between the moral and the material world that was "the crown and glory" of science. What, in his sorrow, he failed to see was his pupil's paradoxical desire to preserve it as well. Positivism enabled Darwin to defend God by removing his guiding hand from nature, but only at the price of his sovereignty.[12]

"The Lord God may be subtle," said Albert Einstein, "but he isn't mean." Darwin would have agreed. In his early notebooks he firmly rejected the idea that God would directly create the world in such a way as to make it appear to have evolved. In the 1842 sketch he sought to involve special creationists in the quandary of believing that God had created independently three species of rhinoceros with deceptive signs of relationship. In the *Origin* he asserted, "certainly we ought not to believe that innumerable beings within each great class have been created with plain, but deceptive marks of descent from a single parent," or that horses that showed unusual stripes were thereby resembling not their own parents or ancestors, but other, unrelated, species. If descent with modification were untrue, then the facts of embryology and rudimentary organs were mere tricks. "To admit this view," he warned,

> is . . . to reject a real for an unreal, or at least for an unknown cause. It makes the works of God a mere mockery and deception; I would almost as soon believe with the old and ignorant cosmologists, that fossil shells had

> never lived, but had been created in stone, so as to mock the shells now living on the sea-shore.

If man's origin had been a unique event in natural history, he wrote in *The Descent of Man,* then the "various appearances" linking him with other animals "would be mere empty deceptions; but such an admission is incredible."[13] The rejection of such possibilities was essential to the logic of a science based on laws and the principles of analogy and uniformity. But they were equally unacceptable to Darwin for theological reasons. And yet, when Darwin was asking whether God would have created a world that appeared to have evolved but had not, he might also have asked himself whether God would have created laws that operated to ends that seemed to be designed and were not. Perhaps he did. Certainly his quandary was deep enough. The question of divine honesty, no less than that of divine innocence, was filled with perplexity.

The implication of divine mendacity, however, was no less objectionable to Darwin than divine irrationality or a waste of the divine energy. As already suggested, a persistent theme in his work was the conviction that the history of life on the earth was best explained as the result of the operation of a system of general laws originally "impressed on matter by the Creator."[14] Best scientifically and best theologically, this union was pithily captured in his first transmutation notebook: "Has the Creator since the Cambrian formation gone on creating animals with same general structure.—Miserable limited view."[15] In the "N" notebook (1838), he wrote impatiently:

> We can allow . . . plants, suns, universes, nay whole systems of universes to be governed by laws, but the smallest insect, we wish to be created at once by special act, provided with its instincts, its place in nature, its range, its—etc. etc.—must be a special act, or result of laws. Yet we placidly believe the Astronomer, when he tells us satellites etc. etc. The Savage admires not a steam engine, but a piece of colored glass, is lost in astonishment at the artificer.—Our faculties are more fitted to recognize the wonderful structure of a beetle than a Universe—

Etienne Geoffroy-St. Hilaire, he had remarked a short time before, "says grand idea God giving laws and then leaving all to follow consequences." Theology and science improved together in this "grand idea":

> Astronomers might formerly have said that God ordered each planet to move in its particular destiny. In same manner God orders each animal created with certain form in certain country, but how much more simple and sublime power [*sic*] let attraction act according to certain laws, such are inevitable consequences—let animals be created, then by the fixed laws of generation, such will be their successors. Let the powers of trans-

> portal be such, and so will be the forms of one country to another.—Let geological changes go at such a rate, so will the number and distribution of the species!!

General laws, that is, those determining the means but not the specific ends, were not only favored by the scientific principle of economy of hypothesis, but they also helped Darwin with the aforementioned problem of evil. In answering, in 1866, a letter asking for help in reconciling science and religion, Darwin expressed preference for the idea of "general laws" governing events in nature such as "pain and suffering" as opposed to "the direct intervention of God," adding, however, "I am aware this is not logical with reference to an omniscient Deity."[16]

It is important to remember that in his attacks on special creation Darwin repeatedly focused on "the ordinary view of creation" as the object of his criticism. There were two reasons for this. First, of course, this was the view that, while not formulated as a doctrine, yet subtly influenced and distracted contemporary science.[17] But, secondly, Darwin was not concerned to refute all possible belief in any sort of creation. His intention was to overthrow the "ordinary view" and to render any others methodologically redundant as far as a scientific explanation of species was concerned.[18] But his own view of the beginning of things was, perhaps, not incompatible with something very like the nomothetic form of special creation.

The origin of life was the problem. Historians have quite properly stressed the importance of Darwin's separating the question of the origin of species from the origin of life. This enabled him to work out a theory of descent without bogging down in unanswerable questions.[19] But it is sometimes implied that he was more or less simply indifferent to the problem of life, and had no ideas about it. This is true of his more consciously positivistic moments. He told Victor Carus, for instance, in 1866, that he rejected spontaneous generation, and said: "I expect that at some future time the principle of life will be rendered intelligible, [but] at present it seems to me beyond the confines of science."[20] But his works do contain references to the problem of the origin and nature of life, and his speculations or explanations sometimes suggest a theological dimension.

Just as general laws were "impressed on matter" by God in a creative act, so the first forms of life, whether one or several, may have been thought of by Darwin, in the 1850s, as some sort of initial creation. "If we admit," he wrote in *Natural Selection,*

> *as we must admit,* that some few organic beings *were originally created, which were endowed with a high power of generation & with the capacity for some slight inheritable variability,* then I can see no limit to the wondrous & harmoni-

ous results which in the course of time can be perfected through natural selection. (italics added)

In this strong statement—stronger than any in the *Origin,* I believe—in which Darwin asserts that one *must* postulate that the first forms of life were already organized for reproduction and inheritance, we have what could be divine initiative and law working together to form a complete explanation of life and its history. The point is, admittedly, elusive. In his conscience-stricken retreat from the jibes at believers in the special creation of each species which he had made in the first edition of the *Origin,* Darwin acknowledged that the conundrums surrounding the appearance of the first forms of life were no less severe for those like himself, who "under the present state of science, believe in the creation of a few aboriginal forms, or some one form of life."[21]

How should "creation" here be interpreted? During his early speculations, Darwin had leaned toward spontaneous generation, considering it the "not improbable" explanation of the origin of life. In his mature years, however, he was often doubtful and sometimes oddly hostile to the idea despite its obvious parallel to pangenesis as a positive theory which, however hypothetical, would account for a fact of nature without inviting theological intrusion into science. The same *Athenaeum* reviewer who twitted him on his use of "Pentateuchal terms" in the *Origin* pointed out that his apparently creationist position on the beginning of life violated the principle of uniformity in favor of "an occult miracle"—to which Darwin replied that "he who believes that organic beings have been produced during each geological period from dead matter must believe that the first being thus arose." As to whether this was probable, he went on, there was not "a fact, or a shadow of a fact, supporting the belief" that life could be generated, by any "known forces," from inorganic matter in the absence of "organic compounds." On the contrary, he said, we are better advised "to confess that our ignorance is as profound on the origin of life as on the origin of force or matter." He concluded, "When the advocate of Heterogeny can . . . connect large classes of facts, and not until then, he will have respectful and patient listeners." And, while in 1872 he wrote to Wallace that he expected spontaneous generation to eventually prove to be true, in the last year of his life he wrote to a correspondent, "I have met with no evidence that seems in the least trustworthy, in favor of spontaneous generation." He did say, however, that he expected the problem of life to be solved by "some general law." In its successive revisions, the *Origin* remained consistently skeptical.

While Darwin, as a positivist, clearly would not have entertained the idea of an initial *miraculous* creation of life (even though joking that it was "less difficult to imagine the creation of an asexual cell" than of a more complex organism), a *lawful* divine initiative employing natural means in the creation-by-law tradition was not ruled out by anything he said. There is some evidence that at one time Darwin believed in a radical discontinuity between the first life form or forms and the normal course of life between the generations of its descendents. "We know," he wrote to Harvey in 1860, "not in the least what the first germ of life was, nor have we any fact at all to guide us in our speculations." Analogy and uniformity could not be pushed to the very beginning of things. The origin of life, if not a divine act, was nonetheless a mystery on which science shed no light. Without suggesting that there were no sound scientific reasons to be skeptical of spontaneous generation—for there were, of course—one may see Darwin's persistent unwillingness to endorse spontaneous generation publically—as he had pangenesis, and despite his evident realization of its theoretical use to him—as consistent with his lingering theologically based cosmology.[22] The origin of life, like design, then, may have been one of those twilight areas in which Darwin's eyes were given to double vision.

A corollary of Darwin's belief that God created only by means of general laws was his firm rejection of any kind of anthropomorphic engineering. In discussing the analogy between the eye and the man-made telescope, he commented, "May not this [analogy] be presumptuous? Have we any right to assume that the Creator works by intellectual powers like those of man?" Ridiculing Argyll's expression that God "labored" in designing new species, he wrote sarcastically to Hooker, "In short, God is a man, rather cleverer than us. . . ."[23] Darwin stressed in the *Origin* that by freeing theology from such restricting anthropomorphisms, his theory made possible a higher conception of deity. In the second edition he added,

> a celebrated author and divine [Charles Kingsley] has written to me that "he has gradually learnt to see that it is just as noble a conception of the Deity to believe that He created a few original forms capable of self-development into other and needful forms, as to believe that He required a fresh act of creation to supply the voids caused by the actions of His laws."

It has been suggested that Darwin welcomed this support from Kingsley as a means of overcoming popular opposition to evolution. It seems to me, in view of his theological concern in the *Origin,* that he also welcomed it as a reassurance of his own belief. The idea was not new to him. He had discussed it in the 1844 essay and in the transmutation notebooks. "How far grander," he wrote in the latter,

> than idea from cramped imagination that God created (warring against those very laws he established in all organic nature) the Rhinoceros of Java

> & Sumatra, that since the time of the Silurian he has made a long succession of vile molluscous animals. How beneath the dignity of him, who is supposed to have said let there be light & there was light—whom it has been declared 'he said let there be light & there was light'—bad taste.[24]

Such cosmic muck-grubbing was not only bad taste; for Darwin, it was very bad theology.

In the well-known passage that concludes *The Variation of Animals and Plants under Domestication* (1868), Darwin objected again to trivializing or slandering the Creator by making him a party to "grotesque" pigeon-fancying and "brutal" bull-baiting. His argument here was, of course, anti-teleological and positivistic; but it was theological as well. One could argue that each variation and subsequent development in nature as well as among domestic animals was designed and that God was simply indifferent to the waste and pain that resulted. This was logically unexceptionable, but Darwin found it morally and theologically unbelievable. By 1868, though, Darwin, perhaps driven by his theological "muddle," was entering his most untheological period, a time from the mid-1860s to the late 1870s when agnosticism was his dominant conscious attitude on religious questions. Consequently, a tone of theological perplexity was a fitting one on which to end the volume. In his closing sentences Darwin presents us with the quandary that he himself had never resolved: if God is "omnipotent and omniscient" then it is hard to see why he is not also irrational and even immoral in producing superfluous laws of nature and waste of life. "Thus we are brought face to face with a difficulty as insoluble as is that of free will and predestination." Asa Gray had accused him of self-contradiction in believing that the laws of nature are designed but the ends that they accomplish are not. Perhaps Darwin came to agree: it was better to give up the idea of God as active in nature than to believe in one who could will such horrors as nature revealed.[25] In the final analysis, Darwin found God's relation to the world inexplicable; and a positive science, one that shut God out completely, was the only science that achieved intellectual coherence and moral acceptability.

8 The Question of Darwin's Theism

It is a not uncommon idea among historians that the theism or, to be more exact, the references to a Creator (the two are sometimes identified) in Darwin's works was put there to turn aside anticipated public wrath over his evolutionary views. The strongest and most frequently cited piece of evidence supporting this interpretation is a letter written to J. D. Hooker on 29 March 1863, discussing the origin of life. "I have," Darwin said,

> long regretted that I truckled to public opinion, and used the Pentateuchal term of creation, by which I really meant "appeared" by some wholly unknown process. It is mere rubbish, thinking at present of the origin of life; one might as well think of the origin of matter.

Inspired by this letter, some writers have questioned the sincerity of Darwin's belief—or professed belief—in a theistic origin of the world, life, and the laws of nature (if not of species). More than one scholar has concluded that the use of the term creation in the *Origin* was merely a sop to the public. Taken alone, the letter seems conclusive in support of that position. But, if the letter is seen in its proper context, this interpretation is much less convincing. Indeed, and in view of the evidence in the preceding pages of this book, it seems to me that such an interpretation requires subscribing to the improbable ad hoc hypothesis that Darwin not only deceived his public in his book, but that he inexplicably deceived his close friends in his correspondence and even himself in his manuscripts and notebooks for over twenty years.[1] Economy of hypothesis, as well as common sense, rebels at such an idea. Bearing in mind not only this, but also the evident integral function of theology in the structure of the *Origin* and his continuing personal reluctance to break with theism, I think there can be no question of Darwin's general sincerity in his theism in 1859. I suspect that the Hooker letter is to be accounted for on two grounds: Darwin's growing agnosticism during the 1860s and his lifelong opposition to biblicism in science. He despised caviling to orthodoxy,[2] and had, he felt, given the special creationists too much comfort and had hurt the cause of positive biology if not of evolution proper. If Darwin was a theist in 1859, there can be no reason to reject out of hand the theism of the *Origin* as protective deception. The Hooker letter, coming four years later, need not even be relevant to that point. I believe, however, that it is relevant and also that it has been misinterpreted by those who accept the "truckling" argument.

There can be no real doubt as to Darwin's theism during the years that he prepared for and wrote the *Origin.* Aside from the strong evidence in his writing, he tells us in his *Autobiography* that the need for postulating an intelligent First Cause as initiating the universe—a belief implied in the theological arguments in the *Origin*—"was strong in my mind about the time, as far as I can remember, when I wrote the *Origin of Species.*" When Dr. Pusey seemed to accuse him of having written the *Origin* as an attack on religion and not as science, Darwin replied indignantly that Pusey was "mistaken in imagining that I wrote the 'Origin' with any relation whatever to theology" (not exactly the case, as we've seen), and that "when I was collecting facts for the 'Origin,' my belief in what is called a personal God was as firm as that of Dr. Pusey himself."[3] The phrase "collecting facts" is chronologically indefinite, but in view of Darwin's concern to refute the charge of atheism directed at the book itself he possibly meant it to include the entire period prior to 1856, when he began writing *Natural Selection,* if not beyond. On the other hand, he might have intended it to refer to the five-year period of alleged Baconian fact-gathering prior to 1842 mentioned in the first chapter of the *Origin*. At the very least, this narrower interpretation would appear to undermine the frequent assertion that Darwin was free of theism during the early years of his theory-building. On Darwin's own testimony, then, as well as on the evidence of his work, one should not hastily dismiss the theistic and even the creationist statements in his writing up to 1859 merely as sops to religious prejudice or as empty bows to convention.

What, then, is the significance of the Hooker letter? The key in the letter is the term "Pentateuchal." That implied creation in the biblical sense of a divine miracle, and as a positivist Darwin of course could not sanction such an interpretation. In a letter to Huxley, written only weeks after the publication of the *Origin,* he spoke of his reference to all life having descended from "one primordial *created* form" as meaning "only that we know nothing as yet [of] how life originates."[4] Francis Darwin refers the passage quoted by his father to p. 484 (p. 753 in the variorum edition) of the *Origin,* which reads slightly but significantly differently:

> Therefore I should infer from analogy that probably all the organic beings which have ever lived on this earth have descended from some one primordial form, *into which life was first breathed.* (italics added)

This was one of the two similar passages to which Darwin later added the phrase "by the Creator," an example of "truckling" perhaps even better known than the Hooker letter. In Darwin's mind, in this case, the term "created" and the biblical image of God breathing life into the newly created Adam were closely associated.[5] I would suggest, then, that it was his ill-considered support of a religious form of special creation by the use of a biblical metaphor that he himself had long been accustomed to employing and not an aversion to theism or even, perhaps, to an initial divine creation by some unknown means, that was the source of his regret and the object of his remark.[6] Along this line, it is suggestive that when militant atheists Edward Aveling and Ludwig Büchner reproached Darwin, during their 1881 visit to Down, with inconsistency in speaking of a Creator breathing life into primordial forms after his fully naturalistic exposition of evolution, he did not defend his action pragmatically or speak of truckling. In fact, he gave no reply at all but was "silent and thoughtful for a space."[7] In any case, these letters are misused when they are employed to support the idea that in 1859 Darwin was a total secularist who concealed his true views behind a facade of religious language.

The presence of theological elements in the *Origin* and the allied problem of the relationship between Darwin's religious views and his science have been obscured by some scholars' tendency to deal too simply with the question of the nature and development of his religion. Not a few writers, loosely following the *Autobiography,* have asserted that Darwin gave up religion at an early age and was henceforth an agnostic.[8] A common mistake in this appraisal is to confuse religion with orthodox Christianity; a second error is to misassign Darwin's later agnosticism to the period prior to 1859.

Darwin's religious disillusionment, as recorded in the *Autobiography,* is familiar. He remembered that he had been conventionally orthodox during the *Beagle* voyage but had come to doubt the full truth of the Old Testament containing as it did a "false history of the world, with the Tower of Babel, the rainbow as a sign, etc. etc." Far worse than any of this was the depiction of God as "a revengeful tyrant." These doubts led Darwin to question the validity of Christianity as a revelation. His growing interest in science led him to think miracles impossible. The contradictions in the Gospels and their not being convincing as eyewitness accounts, coupled with the evident superstitions of the age which they reflected, led him to give up Christianity as revealed. Its spread in the ancient world (a favorite argument in apologetics) meant nothing to him as many false religions had also spread. But, he said, he did not give up his orthodoxy willingly. He daydreamed of evidence that would prove Christianity true but found it harder and harder to imagine arguments that would satisfy him. Hence, in a curious and inverted way,

Darwin discredited Christianity by the same process that he was to use in the 1830s to build scientific theories: hypothetical models of proof. Thus he wrote, "disbelief crept over me at a very slow rate, but was at last complete." He had no doubts or regrets about his loss of faith and was glad enough to get rid of the "damnable doctrine" of the unjust condemnation of unbelievers to hell.[9] The important thing to note at this point is that the break Darwin recounts here is clearly a break with Christianity and not with theism.

That young Darwin's devotion to Christianity was ever very great may be doubted. Sandra Herbert is certainly right to point out that from his Cambridge days his real god was science and that his proposed clerical career was little more than a matter of convenience and filial obedience. He had little piety or deep theological interest to back it up. He did, however, seem to think of religious belief as largely a matter of the intellect.[10] This was not without importance in regard to his theism. On the *Beagle* he showed very little interest in religious worship itself but rather attended church, on board ship or off, for the satisfaction of curiosity or as a diversion. His resolution to read the Greek testament on Sundays seems to have been made with little expectation of realization. He defended the Church of England to some "very pretty Signoritas" in Chile in a spirit of light banter rather than conviction. As late as 1861, he revealed his lack of expertise in biblical matters by replying to the information that the notorious 4,004 B.C. date for the creation of the world was Archbishop Ussher's and not Moses' by laughing, "How curious about the Bible! I declare I had fancied that the date was somehow in the Bible."[11] Against this, of course, must be set his considerable and serious youthful interest in natural theology and his capacity for the spontaneous (although, through the influence of Humboldt, not unanticipated) raptures, the "higher feelings of wonder, astonishment, and devotion, which fill and elevate the mind," which possessed him in the rain forests of South America.[12] Young Darwin was not insensitive to the needs of the spirit or to the allurements of metaphysics, but it was in science and not in conventional religion that they were fulfilled.

In such ways then, Darwin's belief in Christianity slowly vanished. When did it happen? A firm date is impossible to establish. The process was apparent to his wife Emma shortly after their marriage in 1839 and seems to have been well advanced. She wrote her new husband a letter in which she warned him against applying improper tests to religious questions:

> It seems to me also that the line of your pursuits may have led you to view chiefly the difficulties on one side, and that you have not had time to consider and study the chain of difficulties on the other, but I believe you do not consider your opinion as formed. May not the habit of scientific pursuits of believing nothing until it is proved, influence your mind too

> much in other things which cannot be proved in the same way, and which if true are likely to be above our comprehension?

That the falling away that concerned Emma was from Christianity and not theism is proved by her reference later in her letter to the dangers involved in "giving up revelation" and by her own deep faith in the Christian message.[13] In keeping with Victorian piety, her husband's disaffection from the true faith would have grieved her hardly less than if he had avowed complete disbelief in God. Pulling Darwin away from Christianity were not only his difficulties with the Bible and his work in science but also the example of his brother Erasmus (whom Emma mentioned as a bad influence) and Charles Lyell. The geologist was not only shaping young Darwin's scientific ideas at this time, but probably his religious ones as well. Darwin remembered Lyell long after as "thoroughly liberal in his religious beliefs or rather disbeliefs" but as "a strong theist."[14] The Pusey letter, and Darwin's own testimony in his *Autobiography* (already cited), indicate that he, too, was a theist through 1859 and had a strong belief in a "personal God" during at least part of that time. In his autobiography, moreover, he discusses these matters in such a way as to indicate that he did not identify theism with Christianity.[15] In answering Francis Galton's questionnaire in 1873 he wrote, "I gave up *common* religious belief almost independently from my own reflections" (italics added). This must mean Christianity, but no date is given. In 1881, he told Edward Aveling that, because of preoccupation with scientific concerns, "I never gave up Christianity until I was forty years of age" and then because "it is not supported by evidence."[16] This would make 1850 the earliest possible date of a complete break with Christianity. What evidence there is, then, does not yield a very clear picture. This much, however, seems certain. As Darwin says in his autobiography, Christianity slipped away from him very slowly. His later recollection was that it was not completely gone until he had entered middle life. As Christianity weakened during the 1830s and 1840s theism took its place, only to itself surrender to agnosticism during the 1860s.

Recently the question of Darwin's theism has centered on his supposed conversion to materialism during the course of his earliest evolutionary speculations.[17] The issue is not so easily resolved as might at first appear, for materialism and theism are not incompatible. Was Darwin a materialist during the pre-*Origin* years, and, if so, what kind of materialist was he?

There is little doubt that the answer to the first query must be affirmative. The transmutation notebooks are filled with jottings and speculations the meaning of which is unmistakable. Some of these have become almost as familiar as the "truckling" passages in the *Origin:* "thought (or desires more properly) being hereditary it is difficult to imagine it anything but structure of brain"; and, even more pointedly, "Love of deity effect of organization, oh you materialist! . . . Why is thought being a secretion of brain, more wonderful than gravity a property of matter? It is our arrogance, our admiration of ourselves."[18] Darwin was well aware that his type of materialism was viewed with hostility by religious opinion in his society, and was mindful of the fate that had overtaken more than one bold speculator. "To avoid stating how far, I believe, in Materialism," he wrote, "say only that emotions, instincts degrees of talent, which are hereditary are so because brain of child resembles stock—(& phrenologists state that brain alters)."[19] But how far did he believe in materialism? It need not have been very far to get into trouble in those days. Even taking his fear into account, it would seem that Darwin was interested in moving in this direction no farther than was necessary to promote his scientific theorizing. In a note to John Abercrombie's *Inquiries concerning the Intellectual Powers and the Investigation of Truth,* he wrote, "By materialism, I mean, merely the intimate connection of kind of thought with form of brain.—Like kind of attraction with nature of element."[20] The delimiting usage of "merely" is important, for it shows that Darwin was not only restraining his use of the concept of materialism, but that he was employing it as a vehicle for pursuing science as uniformity of law: the analogy with physics makes this evident.

I would suggest that not only was Darwin's materialism compatible in his mind with theism, but that it represented no interest in a thoroughgoing atheistic philosophical or metaphysical materialism. It was, rather, a part of his quest for a science of laws and natural causes, a search of which the descent theory was the fruit. The special creation of species was only one instance of a belief in supposed realms beyond the reach of science, a belief that had to be eliminated. Examples from his notebooks show this. After discussing the causes of psychological states, he remarked, "It is an argument for materialism, that cold water brings on suddenly in head, a frame of mind, analogous to those feelings, which may be considered as truly spiritual." More pointed was a comment on the heuristic advantages of materialism: "strong argument for brain bringing thought, & not merely instinct, a separate thing superadded.—we can trace causation of thought.—it is brought within limits of examination.—obeys same laws as other parts of structure." Materialism, as a network of natural cause and effect, was a postulate necessary for the existence of a science of regularity, system, and prediction. Consequently, free will, an element of caprice and chance, fell under his ban. After reflecting on the inheritance of psychological traits, animal behavior, and related problems, Darwin concluded: "thinking over these things, one doubts existence of free will every action determined by hereditary constitution, example of others or teaching of others." A few pages later, he wrote, "I verily believe free will & chance are synonymous,—shake ten thousand grains of sand together & one will be uppermost,—so in thoughts, one will rise according to law." In the same notebook he commented further, "free will is to mind, what chance is to matter," and, significantly, added later the notation, "M. Le Compte." In notebook "N" he wrote,

> To study Metaphysics, as they have always been studied appears to me to be like puzzling at astronomy without mechanics.—Experience shows the problem of mind cannot be solved by attacking the citadel itself.—The mind is function of body.—we must bring some *stable* foundation to argue from.[21]

Darwin's materialism, it would seem, was nothing more than positivism. It committed him, not to a metaphysics of matter in motion as the ultimate reality, but only to a system of naturalistic and lawful science. This system put God as a participant out of the natural world, it is true; but it did not make it inconsistent to deal with the idea of God or to see him as the Creator of the laws of nature. To interpret Darwin's outlook in the notebooks as that of an "utter materialist" or as "a relatively straightforward materialism"[22] is to make his theological statements in the same series of notebooks, as well as in his published writings, anomalous and to create a problem rather than to solve one. How much simpler, and revealing, to admit that the theology was there because it had meaning for him.

Darwin, of course, was aware that such views as he called materialism tended toward atheism. This was, after all, the conventional religious assumption of his time. His old friend Sedgwick, for example, denounced the *Origin* to a friend: "From first to last it is a dish of rank materialism cleverly cooked and served up. . . . And why was this done? For no other solid reason, I am sure except to make us independent of a Creator. . . ." In a sense, he was right. Darwin, like all positivists, was determined to reduce the study of nature and man to explanations framed solely in terms of physical causes. But "stern physical necessity," Sedgwick protested, did not govern man—or even organic nature. Darwin, as a positivist, disagreed, for positive science could think of man in no other terms. But one need not see him embracing atheism as a consequence of this realization. His interest was not in a "thoroughgoing philosophical materialism," but in a system of science.[23]

Others had the same limited view of scientific materialism. John Tyndall saw this method as a way of conceiving scientific problems and of approaching nature, as a combination of factual knowledge and imagination to be used with restraint and a sense of its limitations. It did not give us theories that duplicated reality, but it was as close to reality as we could come at present. Huxley, in 1868, defended the use of materialistic concepts in scientific work, but denounced any derivation therefrom of a materialistic metaphysics as being as baseless as dogmatic theology. Botanist Karl von Nägeli reminded his colleagues that science dealt only with the finite world and that it had no choice but to proceed "in a strictly materialistic manner," but one "must not forget that this true materialism is an empirical and not a philosophical one, and that it is bounded by the same limits as those of the domain upon which it moves." It is well to remember that, even as a practitioner of "as if," Darwin frequently rejected scientific theories that touched on philosophical materialism: his ambivalence about spontaneous generation; his Lamarck-like insistence on the role of individual will and volition in originating the now inherited expressions of emotions in man and in the higher animals; his dissent from Huxley's notion of animal automatonism, all come to mind.[24] For Darwin, in a deep sense, the world of nature described by science was a creation of science, and as others argued at the time and later, theism or (to some minds) even Christianity was not incompatible with a science predicated on the idea of "the reign of law."[25]

Given his theism in 1859, was Darwin a theist throughout the rest of his life? Many scholars, even those who acknowledge the theism of the *Origin* period, agree that he became a conventional agnostic who rejected religion during his last twenty years. A close look at the evidence, however, indicates once more that this is not so clear. The *Autobiography,* which is the most often quoted authority for his agnosticism, is ambiguous on the point, as is much else in the surviving evidence. Darwin laughed when a German phrenologist pronounced his "bump of Reverence developed enough for ten Priests," but he might have taken pause instead and reflected on the symbolic truth of the claim.[26]

There can be little question but that the dominant tone of the discussion of religion in the *Autobiography* is agnostic. Darwin clearly rejected Christianity and virtually all conventional arguments in defense of the existence of God and human immortality. There is, however, one important exception too often overlooked:

> Another source of conviction in the existence of God, connected with the reason and not with the feelings, impresses me as having much more weight. This follows from the extreme difficulty or rather *impossibility* of conceiving this immense and wonderful universe, including man with his capacity of looking far backwards and far into futurity, as the result of blind chance or necessity. *When thus reflecting I feel compelled to look to a First Cause having an intelligent mind in some degree analogous to that of man; and I deserve to be called a theist.* (italics added)

Darwin went on to express doubts as to the reliability of such conclusions because of their possible origin in the course of human evolution or in infant training. Note the addendum:

> This conclusion was strong in my mind about the time, as far as I can remember, when I wrote the *Origin of Species;* and it is since that time that it has very gradually with many fluctuations become weaker. . . .

This could (mistakenly, I think) encourage one to see the entire statement as retrospective. I would suggest, on the contrary, that the verb tense of the passage clearly shows that Darwin was having moments of theistic reflection—weak though they may have been—even at the time of the writing of the *Autobiography* and even while he was calling himself an agnostic. In 1879, three years after writing his autobiography, Darwin acknowledged that his religious beliefs were constantly shifting. He was, he added, an agnostic, "but not always."[27] This dualism, as it may be called, was consistent from the writing of the *Origin* until his death in 1882.

Darwin confessed his "hopelessly muddled theology" to Gray in 1860 and to Hooker in 1870. "I cannot look at the universe as the result of blind chance," he wrote to the latter, even while denying beneficent design in nature. While expressing doubts or uncertainty to Lyell and to D. MacKintosh in 1874 and 1882, respectively, regarding the existence of a *personal* or *conscious* God (again, the adjectives are important), he wrote to Lord Farrer in 1881, as already mentioned, that the universe cannot be conceived to be the result of chance, "that is, without design or purpose." The reader should also recall his revealing conversation with the Duke of Argyll during his last year. It was the apparent charge of personal atheism that prompted his reluctant but angry reply to Dr. Pusey; and Francis Darwin was at some pains to distinguish Aveling's atheism from his father's agnostic beliefs.[28] In 1881 Darwin read William Graham's *The Creed of Science* (1881) which defended evolution and theism together. "It is a very long time," he wrote to the author, "since any book has interested me so much." While disagreeing with Graham on some points (that the laws of nature imply design and his attempts to limit the role of natural selection), Darwin assured Graham that his book "expressed my inward conviction, though far more vividly and clearly than I could have done, that the Universe is not the result of chance." But, as in the *Autobiography* earlier, he wondered what trust should be put in such subjective convictions. What was their source? Were they atavistic? Would one trust the convictions in a monkey's mind, he asked, "if there are any convictions in such a mind?"[29] And yet, in 1880, in a famous letter traditionally assumed to have been written to Karl Marx but now known to have been sent to Edward Aveling, Darwin wrote that, in his opinion, antichristian and antitheistic polemic seldom had any effect on the public and that "freedom of thought," a goal which he apparently endorsed, was best pursued by promoting science and its consequent enlightenment of mankind. He attributed his own public silence on religious questions, in part, to a respect for the sensibilities of his family.[30]

Darwin's supposedly irenic refusal to engage in the public conflict between science and religion, or, more accurately, between science and orthodox Christianity, has misled some into thinking that he was above the anticlerical animus sometimes felt by other men of science.[31] His abstinence from the type of brawling made famous by Huxley was personal and professional in origin. Such activity would not, in his judgment, advance science and would needlessly wound individuals. In addition, as he wrote to Wallace, "I hate controversy chiefly, perhaps, because I do it badly. . . . " While it is true that his desire for peace led him, on occasion, to fend off inquiries with socially appropriate answers—when Lord Tennyson asked him: "Your theory of Evolution does not make against Christianity?" Darwin is said to have answered, "No, certainly not"—this was, perhaps, more a matter of tactics than a reflection of his true feelings. The tradition of Darwinian reasonableness handed down by the vicar at Down, J. Brodie Innes—"You are . . . a theologian, I am a naturalist, the lines are separate. I endeavor to discover facts without considering what is said in the Book of Genesis. I do not attack Moses, and I think Moses can take care of himself"—is remarkably close to Innes's own sentiments and was contradicted by Darwin himself. In commenting on Dr. Pusey's doctrine of the separation of science and Christianity, Darwin wrote to his vicar, "I hardly see how religion & science can be kept as distant as his reviews [*sic*], as geology has to tell the history of the Earth & Biology that of Man"; however, he added, "I most wholly agree with you that this is no reason why the disciples of either school should attack each other with bitterness, though each holding strongly their beliefs." Nor, he went on significantly, "can I remember that I have ever published a word *directly* against religion or the clergy" (italics added).[32]

Acknowledging the inevitability of conflict, Darwin felt antagonism and resentment toward orthodoxy and its spokesmen. His hostility to biblicism in science was only compounded by his suffering at the hands of ecclesiastical critics; and he was quick to see theological bias in his colleagues. Denouncing a critic of Hooker's British Association address in the *Pall Mall Gazette,* he wrote to Hooker in 1868, "it struck me as monstrous [the author saying] that religion did not attack science. . . ." He thanked Huxley and Carpenter for "checking the odium theologicum" of his reviewers, admired Huxley's prowess in strangling theologian-serpents, and attributed hostile scientific evaluations of the *Origin* by Harvey and Wollaston to their theology. Commenting to Hooker on the *Athenaeum* review of the *Origin,* he complained,

> the manner in which he drags in immortality, and sets the priests at me, and leaves me to their mercies, is base. He would, on no account, burn me, but he will get the wood ready, and tell the black beasts how to catch me. . . .

"I have been atrociously abused by my religious countrymen," he wrote to French naturalist Quatrefages. Undoubtedly, the publication of Gray's pamphlet reconciling natural selection and natural theology and the decision to add the motto from Bishop Butler's *Analogy of Religion,* which he printed in the *Origin* along with those earlier from Bacon and Whewell, were acts intended to head off theological criticism and to encourage readers to judge his theory solely by scientific standards. He chided Lyell for caviling on the question of the origin of man in his tenth edition of the *Principles of Geology* (1867) in a manner "too long, or rather superfluous, and too orthodox, except for the beneficed clergy." He was baffled by St. George Jackson Mivart's unfairness in *The Genesis of Species* (1871) and attributed it to his "accursed religious bigotry."[33] There was never any doubt in Darwin's mind as to a conflict between science and religion nor doubts about what was at stake in it for science.

At the last, as the Aveling letter indicates, Darwin was forced by the imperatives of positivism to couple Christianity and theism as obstacles to science. He had fought their influence all of his scientific career. He vanquished one (perhaps not until middle age) and expelled them both from biology; but he was never able, even in his most agnostic moments, to rid himself entirely of the other. His latter-day theism was one of Darwin's "can't helps," to use Oliver Wendell Holmes, Jr.'s expression. It was not one of his consciously committed positions, but nonetheless he could not get rid of it. Despite its covert significance and a possible role as a censor of scientific ideas, it played no creative role in his science, which was his life, but it was, even so, a part of his total vision of reality up until the end.

This ever-present theism, waning after the publication of the *Origin* but never totally absent from his mind, seems to have served some existential function for Darwin. He distrusted his intuition of a divine dimension in the world and firmly rejected any place for it in science. And yet one wonders to what degree it made his science possible. The theology of the *Origin* suggests that at one time, for him, the rationality and moral probity of God underlay the rationality and meaningfulness of science: it was a metaphysical basis, a remnant of natural theology, a survival of the old episteme, that Darwin required but acknowledged only by implication. It assured him that the laws of nature and the universe which they described were a genuine world, a world the reality and honesty of which could not be demonstrated by positivism alone. Those who dwell on Darwin's intellectual and emotional trauma resulting from the discovery of evolution and of man's place in nature as a cause of his illness might consider the extent to which he may have been protected from these consequences by his residual theism and its emotional reassurances.[34]

Epilogue Positivism and the Decline of Special Creation

The interest of the *Origin of Species* to the historian is intensified by the fact that while it was a harbinger of a new positive biology, it was paradoxically also one of the last major theoretical works of science to be significantly dependent on theology for the force of any part of its argument. Darwin was not in contradiction with himself in this. There were, in effect, two Darwins: one had caught the vision of a new method; the other still adhered to the older view that the very possibility of there being such a thing as science was necessarily linked to theism as the source of meaning and rationality in nature. Within this duality we can summarize the reasons for Darwin's attack on special creation. He rejected the creationist doctrine of divine intervention or superintendence because it was philosophically incompatible with the tenets of an emerging positive science with its emphasis on the uniformity of nature, analogy, the regularity of law, and the full sufficiency of physical causes; because its acceptance restrained the advance of science by erecting metaphysical and mysterious barriers against inquiry; and because, if true, it made God responsible for the horrors, wastes, and irrationalities of nature, and the order of nature itself a delusion and a mockery. Darwin's own approach to evolution fell short of complete positivism. Because of the theological elements in his thought, he continued to speculate—how seriously is admittedly a question—on the possibility of the creation of the first form of life and was loath to abandon the universe to the full meaninglessness that a completely positive view of the cosmos entailed. Yet, paradoxically, it was the prior success of positivism in science that assured the victory of evolution in biology. Kuhn's suggestion that innovation in theory must coincide with a period of growing unease surely explains in part why pre-Darwinian evolutionists failed to convince.[1] Naturalists had to become positivists to some degree before they could become even providential evolutionists, that is, they had to give up interventionism—or, at least, become uneasy about it. Earlier theories, weak though they were in facts and lacking a *vera causa,* also lacked a compelling function in the still predominantly creationist science of the time. Darwin, on the other hand, found scientists becoming more and more positivistic, and made them aware of the implications of this for biology. He made them evolutionists; but, ironically, he could not make them selectionists. As Chauncy Wright noted, "It would seem, at first sight, that Mr. Darwin has won a victory, not for himself but for Lamarck."[2] It is sometimes said that Darwin converted the scientific world to evolution by showing them the process by which it had occurred. Yet the uneasy reservations about natural selection among Darwin's contemporaries and the widespread rejection of it from the 1890s to the 1930s suggest that this is too simple a view of the matter. It was more Darwin's insistence on totally natural explanations than on natural selection that won their adherence.

What had transpired in the minds of many naturalists by 1859 was captured with striking simplicity by Charles Lyell in his journal. "No scientific mind," he wrote, "can imagine an Eleph.t [*sic*] adult coming at once out of nothing." Certainly no one of positivist bent was less inclined to dogmatize about what was possible and impossible than Huxley, but such an event was too much for even his open-minded skepticism. "No doubt," he wrote in his reminiscences about the fight for evolution,

> the sudden concurrence of half-a-ton of inorganic molecules into a live rhinoceros is conceivable, and therefore may be possible. But does such an event lie sufficiently within the bounds of probability to justify the belief in its occurrence on the strength of any attainable, or, indeed, imaginable, evidence?

In a real sense, then, the issue turned on miracles. If a miracle is defined as any sort of supernatural breach of the order of nature, initiated by God for his purposes, and existing beyond the reach of science, then a belief in the impossibility of miracles was the litmus test of positivism, and the spread of skepticism about such intervention was an index to the growth of positive attitudes. Wollaston unintentionally stated the argument for the positivists when he protested, in reviewing the *Origin,* that many miracles of creation were no more absurd than one. More than anything else, it was the incredibility of miracles that disclosed the positivistic commitment to a set of epistemic rules defining reality. It was, above all, *miraculous* creation that offended Darwin both scientifically and theologically. It was a desire to somehow keep God in nature without appealing to miracles that exercised the theoretical skills of the providential evolutionists. What had separated the developing positive outlook from the traditional one, from its significant beginnings in the seventeenth century on, was not the idea that nature is predictable—men had known that from neolithic times if not earlier—but that nature must not be unpredictable. "In common with the most ignorant," wrote John Tyndall, the scientific man

> shares the belief that spring will succeed winter, that summer will succeed spring, that autumn will succeed summer, and that winter will succeed autumn. But he knows still further—and this knowledge is essential to his intellectual repose—that this succession, besides being permanent, is, under the circumstances, necessary. . . .[3]

This need for a nature that was predictably certain was the crux of the reaction against miracles that began in natural history with the nomothetic creationists and came to full flower in positivism where it eventually led to the rejection of any divine role in natural processes or in scientific explanations of them.

One of the most curious features of Lyell's journals on the species question is his repeated observation that a new species or a qualitative leap in the order of nature may come in, indeed does come in, without attendance by miracles and wonders.[4] Why was Lyell so fascinated by this problem? Why did he turn to it again and again and seemingly have to drum the idea of such disruption of the order of nature out of his mind? There can be little doubt that this repeated, ritual-like repudiation of supernaturalism represents his commitment to a science of natural causes and his unease at that commitment, just as it indicates the continuing power of older ideas over the minds of some scientists in the 1850s. Though not a narrow positivist, Lyell shared the widespread belief that the continuity of nature was unbroken.

This common idea, held by laymen as well as scientists,[5] that miracles or supernatural interruptions of the order of nature were impossible, was part of the episteme shift represented by the *Origin.* The impossibility of miracles followed from the implicit metaphysics of positivism: the belief that *all* events are part of an inviolable web of natural, even material, causation. For Tyndall such a simple prayer as one for good weather implied a rearranging of the forces of nature on a scale that would be miraculous if carried out. All prayers to interfere with the determined course of nature, he said, were prayers for miracles and, as such, were beyond the limits of the possible. The believer in miracles, he charged,

> assumes that nature, instead of flowing ever onward in the uninterrupted rhythm of cause and effect, is mediately ruled by the free human will. As regards direct action upon natural phenomena, man's wish and will, as expressed in prayer, are confessedly powerless; but prayer is the trigger which liberates the Divine power, and to this extent, if the will be free, man, of course, commands nature.

Stung by being associated with such a primitive mentality, many theologians and scientific sympathizers abandoned the older position, advocated frankly by Sedgwick, that miracles were violations of the laws of nature, and argued instead that miracles, like nature, are lawful. Miracles in this view were not impossible events, but were applications of higher laws of nature over lower laws. While this argument seemed to make miracles intellectually respectable, in that it reconciled belief in them with a belief in the uniformity of nature, it was self-defeating theologically. It gave the game to the positivist—as, indeed, it testified to the cultural power of science—by removing from the idea of a miracle any identifiable sign of divine action. An event, be it ever so strange, that is natural, that is, within the order of nature, does not logically require and, hence, does not testify to, God.[6]

To be sure, not all subscribed to the idea that miracles were simply impossible. Some, like Huxley, followed David Hume in denying that the so-called laws of nature were anything more than human conventions about the perceived regularity of phenomena and, by themselves, were no justification for rejecting a priori the possibility of preternatural events.[7] For these, the occurrence of events out of the accustomed order of nature was not a matter of possibility but one of probability, of evidence. The Duke of Argyll, disposed though he was to argue with Huxley about most things, agreed with him on this, as did William B. Carpenter who, while a theist, was skeptical of most claims for divine intervention. Yet the appeal to evidence was not quite the easy means of resolution that it appeared. "No evidence," wrote W. K. Clifford in his severe essay, "The Ethics of Belief" (1877), ". . . can justify us in believing the truth of a statement that is contrary to, or outside of, the uniformity of nature. . . ." "No evidence," declared John Stuart Mill, "can prove a miracle to any one who did not previously believe in the existence of a Being or beings with supernatural powers" to work miracles. One might have wondered, with Archbishop Manning (as told by R. H. Hutton), why, with all the emphasis on evidence, skeptics like Huxley did not investigate contemporary miracles where eyewitnesses and scientific evaluation were available. Manning meant Roman Catholic Lourdes, of course, and suggested religious prejudice as the reason for the hesitation, but one might have appealed as well to the miracles of India and the Muslim sects, to those of certain Protestant groups, and others, and still have the same problem. Why people do not do things is not easy to discover. To some extent, Huxley took his battles where he found them and, for the most part, he found them in scripture. But even if he had investigated modern miracles thoroughly, he certainly would have found nothing more than a confirmation that the *event* had happened—which, in theory, he was prepared to admit—and no evidence at all that the event was supernaturally caused. "We can never," he wrote, "be in a position to set bounds to the possibilities of nature." And, continued Mill,

> if we do not already believe in supernatural agencies, no miracle can prove to us their existence. The miracle itself, considered merely as an extraordinary fact, may be satisfactorily certified by our senses or by testimony; but nothing can ever prove that it is a miracle: there is still another possible hypothesis, that of its being the result of some unknown natural cause. . . .

Those who appealed to evidence, therefore, misrepresented the question by suggesting that evidence could settle it. But belief about the miraculous, one way or the other, required a prior belief, that is, an epistemic commitment. "The essential question of miracles," said Baden Powell,

> stands quite apart from any consideration of *testimony;* the question would remain the same, if we had the evidence of our own senses to an alleged miracle, that is, to an extraordinary or inexplicable fact. It is not the *mere* fact, but the *cause* or *explanation* of it, which is the point at issue.

Interpretations of such events, he went on, are governed by our preconceptions. A believer in the uniformity of nature would find, in the mere event, no reason to think it a supernatural intervention or miracle.[8]

There were others, Darwin among them, who, unlike Huxley and his group, were inclined to deny on principle that apparently miraculous events happened, but who were also aware of the potentiality in nature beyond the present knowledge of science. Even if such events should prove undeniable, their commitment to the positive perspective enabled them to reject a supernatural causal explanation. Here it was not claimed to be a question of evidence: a primary incredulity reigned instead. Darwin's reaction to the vogue for spiritualism which swept up Wallace and other scientists in the 1860s and 1870s was analogous to the positivist view of miracles: "The Lord have mercy on us all, if we have to believe in such rubbish." Regarding the testimony of physicist William Crookes, a leading psychical researcher, he wrote: "I cannot disbelieve Mr. Cooke's statement nor can I believe in his result . . .," and then suggested analogies with physical laws. Elsewhere he protested, "to my mind an enormous weight of evidence would be requisite to make one believe in anything beyond mere trickery." To his vicar, he wrote in 1860: "I must say that I am a complete skeptic about the powers at work,—curious as your stories are. What stories one hears about the spirit-tapping now-a-days—the old saying to believe nothing one hears and only half of what one sees is a golden rule."[9]

Once admit such a state of affairs, whether miracles or spirits, and it would be a constant threat to scientific generalization and prediction. Such a world would defy scientific investigation. The supernatural, the immaterial, another order of being penetrating nature but not a part of it, made science (as the positivist viewed science) impossible. That interpretation of the uniformity of nature that Asa Gray denied found any sanction in science was exactly what the positivist insisted on: "that every natural effect is, and has ever been, preceded by natural causes."[10] The positive position was not philosophically unassailable and there were those who assailed it. But such arguments, whatever their merits, were helpless to stem the flood tide of positivist conviction among working scientists. "Science," Tyndall told the British Association at Belfast in 1874, "demands the radical extirpation of caprice and the absolute reliance upon law in nature." For Huxley, "the fundamental axiom of scientific thought is that there is not, never has been, and never will be, any disorder in nature." Any exception "would be an act of self-destruction on the part of science." Joseph LeConte believed in evolution despite what he took to be the adverse verdict of geology because regularly occurring "secondary causes and processes" were all that science knew, and that meant evolution. Natural science, wrote Nägeli, should avoid "immaterial principles"; they were superfluous, because "everything can be explained in a natural way." Darwin concurred: "Spiritual powers cannot be compared or classified by the naturalist." "No inductive inquirer," wrote Baden Powell,

> can bring himself to believe in the existence of any *real hiatus* in the continuity of physical laws in past eras more than in the existing order of things; or to imagine that changes, however seemingly abrupt, can have been brought about except by the gradual agency of some regular causes. On such principles the whole superstructure of rational geology entirely reposes: to deny them in any instance would be to endanger all science.

"There can no longer be any more doubt," wrote George Romanes triumphantly in 1878,

> that the existence of a God is wholly unnecessary to explain any of the phenomena of the universe, than there is doubt that if I leave go of my pen it will fall upon the table. . . . the knowledge that a Deity is superfluous as an explanation of anything, being grounded on the doctrine of the persistence of force, is grounded on an *a priori* necessity of reason—i.e., if the fact were not so, our science, our thought, our very existence itself, would be scientifically impossible.

Canadian antievolutionist John William Dawson came closer to the mark than he usually did when he accused his fellow scientists of rejecting direct creation because they were corrupted by materialism and would not consider spiritual forces as a part of nature. "Views like [Darwin's] rest," agreed William Hopkins, "in fact, on no demonstrative foundation, but . . . on *a priori* considerations, and on what appears to us a restricted, instead of an enlarged view of the physical causes, operations, and phenomena which constitute what we term Nature."[11]

The positivists' rejection of miracles, then, was a position based not on experience (save the experienced fact that belief in many alleged miracles had been the consequence of ignorance),[12] for no experienced uniformity could contravene an event that by definition violated uniformity, but on the theoretical needs of their science. The essence of positive science was predictability: caprice had no place in its cosmos. Those who ruled out miracles as contrary to the laws of nature were correctly taxed with begging the question: such axioms always do. But when asked to establish the probability of this or that specific miracle (either biblical or biological), the orthodox could do little but beg the question themselves by appealing to scripture or faith or theology, or to the ignorance of science. In this mutual begging of questions, we see the disparate nature of two conflicting epistemes.

Special creation, then, understood as a *direct* divine intervention in nature, was as much a victim of an increasing awareness of the changing requirements of scientific thinking as of anything else. As scientists, through their work, became more systematically reflective about what they were doing, they became less able to take seriously the miracles or mysterious divine initiatives of the special creationists and their biblicist recasting of ancient stories and religious symbols as scientific accounts of the development of nature. It was a case, as Foucault has said in another connection, of "the stark impossibility of thinking *that.*" The primary change had not been in speciation theory but in beliefs about the nature of science. This change, once accomplished, simply nullified special creation as a scientific idea.[13]

As suggested earlier, an understanding of the nineteenth-century episteme shift among men of science revitalizes the much abused concept of the conflict of science and religion and permits a deeper comprehension of it. The orthodox claim that true science was not hostile to faith meant that a science formulated in terms of the older episteme was not. But this science was unacceptable to the positivist. Its epistemic assumptions were, to him, both functionless—or even harmful—and incredible. That church officials could no longer imprison or suppress scientists, that many scientists were themselves devoted to religion, is true but not to the point. The threat lay in this: the very existence of a rival science or of an alternate mode of knowledge was intolerable to the positivist. His emotional attachment to science as the pursuit of truth,[14] as a lifelong commitment, as a profession, his awareness of the historic novelty of his situation, made him intolerant of all other claims to scientific knowledge. Anyone not of his tribe was a charlatan, an imposter. Historians who have denounced the rhetoric of Huxley and Draper as not giving a true picture of the relations between science and religion at that time are right in a narrow sense. But they are right at the cost of overlooking the major changes that activated Huxley and Draper. These changes were both intellectual and institutional. It was not enough to drive out the old ideas. Their advocates had to be driven out of the scientific community as well. It is in this way that we should understand the personal conflicts, the battles within scientific societies, the purging of parson-naturalists and other amateurs. In order for the world to be made safe for positive science, its practitioners had to occupy the seats of power as well as win the war of ideas. Both were necessary to the establishment of a new scientific orthodoxy. Accordingly, it would be a serious error to see the conflict over evolution as primarily one of generational or professional strife in which the question of evolution was merely the occasion for the battle. On the contrary, generational and professional hostility was necessary *because* of evolution or, more accurately, because of positivism and the distinctive approach to scientific thinking that it involved.

It is important to notice that it was not necessary for a scientist to renounce religion in order to be a member in good standing of the new order. Simple theism, such as Darwin possessed in 1859, interfered little with the practice of science because it had no doctrines that prescribed beliefs about the world. The more complex the theology, the greater was the potential for interference. The problem, then, was not theism, but positive theological content. Scientists who were theists could also be positivists. Those who were orthodox usually became more liberal in their theological views as they drew closer to positive science. The shift from one episteme to another required not the surrender of religion as such, but rather its replacement by positivism as the epistemological standard in science. And this eventually took God out of nature (if not out of reality) as effectively as atheism.[15] That religion could continue under such terms often concealed from participants what had actually occurred. Nor were they the only ones deceived. In the new episteme reality was always an inference. Men would never be able to claim certainty for their beliefs while they continued within its boundaries. Popularizers of the new science who spread a gospel of metaphysical materialism based on science's supposed certain authority appreciated the real significance of what had happened as little as did the theologians who thought successful accommodation of a divinely revealed religion to the new science was a simple matter of shedding a few antiquated superstitions.

As suggested at the beginning of this book, an intriguing question in all this is how and why one passes from one episteme to another. For some, once the old belief was altered, there was no compromise. For them, accommodations, the updating of religion to make it agree with contemporary science, was as meaningless a manipulation of discredited symbols and bankrupt ideas as was the old science. A new world had opened for them. Men, however, are not converted to new epistemes in an instant. They move, live, work into them at varying speeds and in stages. It would be foolish to believe that scientists became evolutionists by reason of a sudden envelopment in a marvelous epistemic cloud that seemingly settled from heaven. Men do not drop one episteme and pick up another for no cause. Long and often anguished battles with facts and theory preceded most shifts in allegiance. Out of this dialectic struggle within each naturalist emerged a more adequate understanding of nature that was the product of a new way of doing science, and the new science was brought about by the failure of the old to accommodate new facts and to answer new questions. While it is true that making the assumptions of positive science necessarily ruled out any idea of special creation, those assumptions did not necessarily require a belief in evolution. There were logically possible, although not very probable, alternatives to special creation which were compatible with those assumptions. Lyell's mysterious creative force, if stripped of its Creator, could have been one of these. Others, more fanciful, come to mind. And nescience, however unproductive and contrary to the internal logic of positivism, was always an option, if not a solution. But there was a practical consensus, so strong that it was tantamount to necessity, that evolution of some kind, that is, speciation through a process of physical descent, was the only alternative to special creation. Even within the positive episteme, however, evolution required evidence and any particular formula such as natural selection especially did. But, as Kuhn has pointed out, a change of paradigm is not merely a function of logic or the observation of nature. These things, in fact, require the paradigm for their full exploitation.[16] And the paradigm, evolution, rested on a prior epistemic change. It was this alteration of episteme growing out of the practice of science that accounts for the fact, noted by Huxley and others, that evolution was accepted by naturalists long before it could be said to have been proven. A new episteme may, it is true, merely be a change of illusion, but it may also provide a real advance in knowledge.

In any event, some men moved faster than others to accept the new perspective. Some were willing to accept evolution, even natural selection, if they could believe God was behind it. To others, those farther into the new episteme, this divine activism was redundant, if not absurd. "Belief in a creation of life," wrote the German Darwinian Oscar Schmidt, "is incompatible with investigation of it."[17] This could stand as a motto for the new outlook. For those having it, as Foucault wrote, "things are no longer perceived, described, expressed, characterized, classified, and known in the same way."[18] Old arguments that once seemed axiomatic lose their force, or even their comprehensibility. Old ideas may live on briefly as objects of attack, but not as real options. After 1859, special creation entered this last stage of a live idea. It existed only to be refuted. After a while, it was simply not there anymore. "I am deeply convinced," Darwin wrote to Lyell shortly before the *Origin* was published, "that it is absolutely necessary to go the whole vast length, or stick to the creation of each separate species."[19] More fully than he understood, there was no permanently stable midpoint where one could safely rest and have the best of both worlds—an unwelcome realization that, at the last, Darwin himself was reluctant to accept.

The end of special creation as a plausible scientific idea was the beginning of the end for significant obstacles to a completely positive biology. The direction had been set early in the nineteenth century by the creation of biology itself, as opposed to the earlier natural history. Interest shifted from taxonomy to analysis, from the observation, description, and cataloging of living things to a concern with inquiring into the why of their constitution and interrelationships.[20] Agassiz's descriptive idealism with its reliance on theology for explanations would no longer do. The *Origin of Species* played a central part in opening up a new biology for both scientists and laymen.[21] In convincing men to be evolutionists, Darwin helped to teach them to explain biology in terms of lawful natural causes and to use explanatory theories which were capable of formulation within the conventions of positive science. Despite the temporary popularity of providential evolution, and the resurgence of vitalism, the defenses were breached and eventual success assured. In less than a century, biology had taken its place within the dominant philosophy of science. So complete was the acceptance of positivism that even the opponents of natural selection who appeared at the turn of the century and who, like the providential evolutionists, advocated mutationist theories of evolution avoided the appeals to designed evolution and its associated teleology that had been characteristic of Darwin's earlier adversaries. Huxley's pained cry of 1860, "Shall Biology alone remain out of harmony with her sister sciences?" had been answered in the negative.[22]

Darwin's own career should be no less instructive to the historian than his book was to his contemporaries. The story of his intellectual development is that of a man's outlook evolving, if one will, from something akin to Comte's theological state to his positive stage. In the *Origin,* Darwin used both views alternately and even simultaneously. Gradually, after the *Origin* was written, the positive orientation moved ahead to dominate his scientific work. Darwin's progress toward positivism in science must have been duplicated in other scientists and scientifically inspired laymen: a dialectical movement through particular scientific work to a general acceptance of positivism as a tool for such work and as a world view. His life is a model example of how one episteme displaces another.

Notes

Preface

1. The extent and richness of the scholarship may be learned from John C. Greene's essay, "Reflections on the Progress of Darwin Studies," *Journal of the History of Biology* 8 (1975): 243–74.

Chapter 1

1. See Traian Stoianovich, *French Historical Method: The Annales Paradigm* (Ithaca: Cornell University Press, 1976), p. 211.

2. Michel Foucault, *The Order of Things: An Archaeology of the Human Sciences* (London: Tavistock, 1970), p. ix. The original edition is *Les mots et les choses: une archéologie des sciences humaines* (Paris: Gallimard, 1966). All citations are to the English edition.

3. Michel Foucault, *The Archaeology of Knowledge,* trans. A. M. Sheridan Smith (New York: Harper and Row, 1972), pp. 138–39, 147. The original edition was *L'archéologie du savoir* (Paris: Gallimard, 1969). All citations are to the English edition. For an application closer to Foucault's own approach, see Karl M. Figlio, "The Metaphor of Organization: An Historiographical Perspective on the Bio-medical Sciences of the Early Nineteenth Century," *History of Science,* 14 (1976): 17–53.

4. Ibid., pp. xxii, 158. In *Archaeology,* Foucault neglects the term "episteme" for "positivity" or "historical a priori." A "positivity" is the "rules" under which "a discursive practice may form groups of objects, enunciations, concepts, or theoretical choices." Positivities "form the precondition of what is later revealed and which later functions as an item of knowledge or an illusion, an accepted truth or an exposed error" (pp. 181–82). The episteme, however, returns (pp. 191–92) apparently as a more inclusive term for the study of "positivities" in a certain period of time.

5. Thomas S. Kuhn, *The Structure of Scientific Revolutions,* 2d ed. (Chicago: University of Chicago Press, 1970), pp. viii, 10, 85, 149. In "Second Thoughts on Paradigms" in *The Structure of Scientific Theories,* ed. Frederick Suppe (Urbana: University of Illinois Press, 1974), Kuhn substitutes "disciplinary matrix" and "exemplar" as designations of the broader and narrower meanings (p. 463).

6. Foucault, *Order,* p. xi.

7. In *L'archéologie du savoir* Foucault uses "savoir" and "connaissance" in a similar way. See *Archaeology,* p. 15. My use, below, of the origin of new species through natural descent as a paradigm does not parallel Kuhn's usage exactly because it conveyed no classic exemplar or theory; rather, it provided a methodological imperative, i.e., this was the means that was mandated within the positive episteme. Darwin's candidate for "exemplar," natural selection, did not attain paradigm status until the 1930s, if then.

8. Michael Ruse, "The Revolution in Biology, *Theoria,* 35 (1970): 13–22; idem, "Two Biological Revolutions," *Dialectia,* 25 (1971): 17–38; John C. Greene, "The Kuhnian Paradigm and the Darwinian Revolution," in *Perspectives in the History of Science and Technology,* ed. Duane H. D. Roller (Norman: University of Oklahoma Press, 1971), pp. 3–25; Michael

Ghiselin, "The Individual in the Darwinian Revolution," *New Literary History,* 3 (1971): 113–34; Ernst Mayr, *Evolution and the Diversity of Life: Selected Essays* (Cambridge: Harvard University Press, 1976), pp. 277–94. Ruse's analysis of the crisis in natural history, independently and to a slight degree, resembles mine.

9. Foucault, *Order,* p. 168. Foucault seems to have decided not to mean this as literally as it sounds. See *Archaeology,* pp. 148, 158–59, 178–81. Conflict may arise within a single episteme, however, as part of the process of working out its implications. See *Order,* pp. 74–75.

10. I use the term "creationism" for the older episteme with some reluctance because of its common, but misleading, equation with miracle. It has been impossible, also, to avoid using the word "create" occasionally in the more general sense of "originate." I can only hope that by warning the reader confusion may be avoided.

11. David L. Hull in his *Darwin and His Critics: The Reception of Darwin's Theory of Evolution by the Scientific Community* (Cambridge: Harvard University Press, 1973), p. 67, has suggested "neoplatonism" as a better designation than "idealism" for this view since the latter means "that only ideas exist." But, since neoplatonism in its historical manifestations was as much an esoteric religion as a philosophy, that term seems hardly more suitable. In truth, the scientific outlook, in both epistemes, resists pigeonholing in academic philosophic categories: idealist/materialist, empiricist/rationalist, nominalist/realist. Such terms are only approximate, not definitive. Practicing naturalists were not so systematic in their thinking as such designations would imply. In any era, the practice of science is seldom identical with its philosophy. In this case, it seems clear that the ideas of Louis Agassiz, Richard Owen, and other "idealists" about species were more grounded in a theology of creation than in an formal epistemology. Therefore, while acknowledging the technical point Hull raises, I will continue the customary usage as defined. There is no suggestion that idealists throughout history have all been "creationists."

12. Foucault, *Order,* pp. 127–28, 150–58, 160–61, 265, 269, 273, 281. In my usage, the terms "natural history" and "biology," unlike Foucault's, have no special meaning beyond the intended one that biology was transformed into a positive science under the new episteme.

13. Among commentators on Kuhn and Foucault, see Dudley Shapere, "The Structure of Scientific Revolutions," *Philosophic Review,* 73 (1964): 391–92; E. S. Shaffer, "The Archaeology of Michel Foucault," *Studies in History and Philosophy of Science,* 7 (1976): 274; Israel Scheffler, *Science and Subjectivity* (Indianapolis: Bobbs-Merrill, 1967), pp. 15–19, 76–89; Derek L. Phillips, "Paradigms and Incommensurability," *Theory and Society,* 2 (1975): 37–61; Stephen Toulmin, *Human Understanding* (Oxford: Oxford University Press, 1972), pp. 101–02; Roger Trigg, *Reason and Commitment* (New York: Oxford University Press, 1973), pp. 99–118; David Leary, "Michel Foucault, an Historian of the *Sciences Humaines,*" *Journal of the History of the Behavioral Sciences,* 12 (1976): 286–93; various writers in Imre Lakatos and Alan Musgrave, eds., *Criticism and the Growth of Knowledge* (Cambridge: Cambridge University Press, 1970); Hull, *Darwin and His Critics,* pp. 451–55. Kuhn has denied the extreme view attributed to him: see "Reflections on My Critics" in Lakatos and Musgrave, *Criticism,* pp. 260–64, 267–77, and "Second Thoughts," ibid., pp. 506, 508.

14. On this point, see Shapere, "Structure," p. 393; Ghiselin, "The Individual"; Phillips, "Paradigms," pp. 55–57; Toulmin, *Human Understanding;* Kuhn, "Second Thoughts," pp. 477–78, 512–13; Martin J. S. Rudwick, "Darwin and Glen Roy: A 'Great Failure' in Scientific Method?" *Studies in History and Philosophy of Science,* 5 (1974): 97–185.

15. See Trigg, *Reason and Commitment,* pp. 99–118 for a vigorous denunciation of some of these paradoxes. Bacon wrote: "For the wit and mind of man, if it work upon matter, which is the contemplation of the creatures of God, worketh according to the stuff, and is limited thereby: but if it work upon itself, as the spider worketh in his web, then it is endless and brings forth indeed cobwebs of learning, admirable for the fineness of thread and work, but of no substance or profit" (*The Advancement of Learning,* book 1).

16. Toulmin, *Human Understanding,* pp. 104–5, points out the importance of this in the transition from Newtonian to Einsteinian physics.

17. Sheffler, *Science and Subjectivity,* pp. 79–87; W. F. Cannon, "Charles Lyell, Radical Actualism, and Theory," *British Journal for the History of Science,* 9 (1976): 105–6.

18. Kuhn, *Scientific Revolutions,* pp. 7, 21, 150–58. "Conversions" were not absent in the shift to evolution. See, for example, Alfred Newton's account of his own experience in "The Early Days of Darwinism," *Macmillan's Magazine,* February 1888, pp. 241–49. But, as Newton acknowledges, most of his contemporaries did not follow his pattern. Asa Gray, on the other hand, stressed the practical problem-solving role of evolutionary theory as the key to its success; see *Darwiniana,* ed. A. Hunter Dupree (Cambridge: Harvard University Press, 1963), pp. 195–207. See also Joseph D. Hooker's remarks as recorded by Henry Fawcett, in Hull, *Darwin and His Critics,* p. 290. According to Howard Gruber, Darwin's own experience does not seem to have been that of a "conversion." See Howard E. Gruber and Paul H. Barrett, *Darwin on Man* (New York: E. P. Dutton, 1974), pp. 4, 130–31, 146, 170.

19. Foucault, *Order,* pp. xii–iii, xxiii, 50–51, 217–21, 250–51, 274. Foucault has some idea of workaday experience in changing ideas, but is reluctant to give this element much credit for episteme change as such: see pp. 228, 230–32, 236–39, 246, 252–53, 268, 270–76. See also idem, *Archaeology,* chap. 5, esp. pp. 172–74, 179, 208–9; Shaffer, "Archaeology," p. 275.

20. Illustrative of various definitions are W. M. Simon, *European Positivism in the Nineteenth Century* (Ithaca: Cornell University Press, 1963); N. G. Annan, *The Curious Strength of Positivism in English Political Thought* (Oxford: Oxford University Press, 1959); Ernst Cassirer, *The Problem of Knowledge: Philosophy, Science, and History since Hegel* (New Haven: Yale University Press, 1950): C. C. Gillispie, *The Edge of Objectivity* (Princeton: Princeton University Press, 1960); Nicola Abbagnano, in *Encyclopedia of Philosophy,* s.v. "positivism"; D. G. Charlton, *Positive Thought in France during the Second Empire* (Oxford: Oxford University Press, 1959); Leszek Kolakowski, *The Alienation of Reason: A History of Positivist Thought* (Garden City, N.Y.: Doubleday, 1968); Maurice Mandelbaum, *History, Man and Reason: A Study in Nineteenth-Century Thought* (Baltimore: Johns Hopkins University Press, 1971). Abbagnano, Charlton (pp. 5–6), and Mandelbaum (pp. 10–11) have understandings of positivism most like mine.

21. It has been suggested that "naturalism" might be a better term than "positivism" in this study. On the other hand, conventional historiographical usage, which has often equated "naturalism" with any science based on laws, has tended to obscure the theological ground of beliefs such as "creation by law" and hence has blurred important distinctions (to be discussed below) between that approach to the problem of species and the approach which I call "positivism." Positivism, as defined in this book, while not a perfect choice of term, does serve to avoid that particular confusion of lawful systems of science, whatever other confusions it may invite. Unfortunately, I have found it difficult for stylistic reasons to avoid the occasional use of "naturalism" in some form as a synonym for "positivism." Again, one can only hope that forewarning will offset misunderstanding.

22. Hull notes the growing acceptance of natural explanations among biologists in the nineteenth century as well as a greater self-consciousness about the right way to do science; see *Darwin and His Critics,* pp. 4, 75.

23. *The Conduct of Inquiry: Methodology for Behavioral Science* (Scranton, Pa.: Chandler, 1964), pp. 3–11. For the challenges to contemporary philosophy of science contained in Darwin's theory, see David L. Hull, "Charles Darwin and Nineteenth-Century Philosophies of Science," in *Foundations of Scientific Method: The Nineteenth Century,* ed. R. N. Giere and R. S. Westfall (Bloomington: Indiana University Press, 1973). On the contribution of working naturalists to the intellectual reorientation in biology see Ernst Mayr, "Open Problems in Darwin Research," *Studies in History and Philosophy of Science,* 2 (1971): 279; for paleontologists, see Peter J. Bowler, *Fossils and Progress* (New York: Science History Publications, 1976), p. 112.

24. Michael T. Ghiselin sees this feature as the important thing in Lyell's uniformitarianism; see *The Triumph of the Darwinian Method* (Berkeley: University of California Press, 1969), p. 14.

25. Kuhn, *Scientific Revolutions,* p. 4.

26. Baden Powell, *Essays on the Spirit of the Inductive Philosophy, the Unity of Worlds and the Philosophy of Creation* (London: Longmans, 1855), pp. 359–60, 402, 423, 425.

27. Thomas Henry Huxley, "The Origin of Species," in *Lay Sermons, Addresses and Reviews* (New York: D. Appleton, 1871), pp. 279–82.

28. Charles Lyell, *Principles of Geology,* 11th ed., 2 vols. (New York: D. Appleton, 1873), 2:330 (hereafter cited as *Principles* [1873]).

29. On the important role of Laplace's nebular hypothesis in this regard see Ronald L. Numbers, *Creation by Natural Law* (Seattle: University of Washington Press, 1977).

30. See Owen Chadwick, *The Secularization of the European Mind in the Nineteenth Century* (Cambridge: Cambridge University Press, 1975) on the complexity of the causes of secularism.

31. For the concern of seventeenth-century scientists to avoid strife between religion and science see Richard S. Westfall, *Science and Religion in Seventeenth-Century England* (New Haven: Yale University Press, 1958).

32. For the development of an articulate idealism in pre-Darwinian natural history and allied sciences, see Martin J. S. Rudwick, *The Meaning of Fossils: Episodes in the History of Palaeontology* (London: Macdonald, 1972); Peter J. Bowler, "Darwinism and the Argument from Design: Suggestions for a Reevaluation," *Journal of the History of Biology,* 10 (1977): 29–44.

33. Duke of Argyll, *Inaugural Address . . . on . . . Installation as Chancellor of the University of St. Andrews . . .* (Edinburgh: James Stillie, 1852), pp. 14–16; Wollaston, in Hull, *Darwin and His Critics,* pp. 138–39; Richard Owen, "On the Argument of 'Infirmity' in Mr. Lewes' Review of *The Reign of Law,*" *Fraser's Magazine,* October 1867, pp. 531–33. For the perception of design as a mental set see Vincent C. Kavaloski, "The *Vera Causa* Principle: An Historical Philosophical Study of a Metatheoretical Concept from Newton to Darwin," (Ph.D. diss., University of Chicago, 1974), chap. 8.

34. Adam Sedgwick, *Discourse on the Studies of the University of Cambridge,* 5th ed. (London: Parker, 1850), pp. xi–xii, cxxxvii, 150; *The Life and Letters Adam Sedgwick,* ed. John Willis Clark and T. McHughes, 2 vols. (Cambridge: Cambridge University Press, 1890) 2:357, 360; *Princeton Review* (1856) as quoted in Major L. Wilson, "Paradox Lost: Order and Progress in Evangelical Thought of Mid-Nineteenth-Century America," *Church History,* 44 (1975): 362.

35. William B. Carpenter, *Nature and Man* (London: Kegan Paul, 1888), pp. 365–66, 369, 379–80, 382, 396, 407, 413; Asa Gray, *Letters of Asa Gray,* ed. Jane Loring Gray, 2 vols. (Boston: Houghton, Mifflin, 1893), 2:592.

36. Chadwick, *Secularization,* p. 167.

37. Cope, quoted in Henry Fairfield Osborn, *Cope: Master Naturalist* (Princeton: Princeton University Press, 1931), pp. 536, 560; also [Cope], *American Naturalist* 20 (1886): 708–09; Cope, "The Foundations of Theism," *Monist,* 3 (1893): 623–39; idem, *The Theology of Evolution* (Philadelphia: Arnold, 1887), pp. 28–29, 39.

38. A reexamination of the question of religion among scientists is urged in David B. Wilson, "Victorian Science and Religion," *History of Science,* 15 (1977): 52–67. While Wilson deals mainly with the physical sciences, a full-scale reappraisal is needed. I would suggest that the physical sciences were the least challenging to religion because of the Newtonian accommodation achieved earlier. The materialist challenge of the new science and its attack on design in nature were felt most strongly in the life sciences where the connection with the creationist episteme was most evident. On the science-religion question, see also Frank M. Turner, *Between Science and Religion: The Reaction to Scientific Naturalism in Late Victorian England* (New Haven: Yale University Press, 1974).

39. Thomas Henry Huxley, *The Life and Letters of Thomas Henry Huxley,* ed. Leonard Huxley, 2 vols. (New York: D. Appleton, 1900), 2:9; Carpenter, *Nature and Man,* pp. 239–40; Sebastian Evans, in *Nature,* 3 (1870–71): 362, Charles Darwin, *The Life and Letters of Charles Darwin,* ed. Francis Darwin, 2 vols. (New York: D. Appleton, 1896) 2:252n

(hereafter cited as *DLL*); T. G. Bonney in Hester Pengelly, ed., *A Memoir of William Pengelly, of Torquay, F. R. S., Geologist* . . . (London: John Murray, 1897), p. 317.

40. John Fiske, "Draper's Science and Religion," *Nation*, 28 November 1875, pp. 343–45.

Chapter 2

1. W. F. Cannon, "Darwin's Vision in *On the Origin of Species*," in *The Art of Victorian Prose*, ed. George Levine and William Madden (New York: Oxford University Press, 1968), p. 156. See also Cannon, "The Basis of Darwin's Achievement: A Revaluation," *Victorian Studies*, 5 (1961): 109–34.

2. A good treatment of Darwin's general religious development is Maurice Mandelbaum, "Darwin's Religious Views," *Journal of the History of Ideas*, 19 (1958): 363–78. As will appear, I have some points of difference with the author.

3. Foucault, quoted in Stoianovich, *French Historical Method*, p. 232.

4. Charles Darwin, *The Origin of Species, by Charles Darwin: A Variorum Text*, ed. Morse Peckham (Philadelphia: University of Pennsylvania Press, 1959), p. 750 (hereafter cited as *Origin*).

5. These two categories, and those which follow, are schematic in the sense that, in reality, individuals might fall into more than one category, leave one for another, or move back and forth in the course of their scientific careers. Individual views are offered as illustrative of more widespread attitudes. My rationale for generalizing is this: To show that Darwin was reasonable in his attack on special creation, I need only show that ideas contrary to his were present in the scientific community and were held by prominent and influential naturalists. Assuming that one of the functions of professional leadership is to form as well as to reflect opinion, I further assume that the beliefs of these men represent similar ideas among less articulate naturalists.

6. Charles Lyell, *Principles of Geology*, 3 vols. (London: John Murray, 1830–33), 2:128 (hereafter cited as *Principles* [1830]). Lyell's italics have been omitted.

7. Edward Hitchcock, *Elementary Geology*, 13th ed. (New York: Ivison and Phinney, 1859), p. 334; Sedgwick, *Discourse*, pp. xvii–xxii, ccxiv; idem, *Life and Letters of Sedgwick*, 2:361; Gideon Mantell, quoted in a review of his *Wonders of Geology* in *American Journal of Science*, 39 (1840): 13–14.

8. Hugh Miller, *Footsteps of the Creator, or the Asterolepsis of Stromness*, 11th ed. (Edinburgh: Nimmo, 1869), pp. 293–94; Edward Hitchcock, *The Religion of Geology and Its Connected Sciences*, new ed. (Boston: Phillips, Sampson, 1859), pp. 528, 564–65; idem, *Elementary Geology*, pp. 334–35. Hitchcock, of course, did not regard God as a sort of wizard. For his attempt to establish the existence of a law of miracles, but one unknowable by man, see Stanley M. Guralnick, "Geology and Religion before Darwin: The Case of Edward Hitchcock, Theologian and Geologist (1793–1864)," *Isis*, 63 (1972): 535–37.

Louis Agassiz, *Essay on Classification*, ed. Edward Lurie (Cambridge: Harvard University Press, 1962, p. 15, n. 15; Joseph LeConte, "Lectures on Coal," *Annual Report of the Smithsonian Institution, 1857* (Washington, 1858), p. 168; Louis Agassiz and Augustus A. Gould, *Principles of Zoology* (Boston: Gould, Kendall and Lincoln, 1848), p. 182 and passim.

9. *Principles* (1830), 2:84, 119; Charles Darwin, *Charles Darwin's Diary of the Voyage of H.M.S. "Beagle,"* ed. Nora Barlow (Cambridge: Cambridge University Press), 1934, p. 236 (hereafter cited as *Diary*); Charles Darwin, *Journal and Remarks, 1832–1836*, Narrative of the Surveying Voyages of His Majesty's Ships Adventure and Beagle . . . , vol. 3 (London: H. Colburn, 1839), p. 474 (hereafter cited as *Journal* [1839]); Charles Lyell, *Sir Charles Lyell's Scientific Journals on the Species Question*, ed. Leonard G. Wilson (New Haven: Yale University Press, 1970), p. 198 (hereafter cited as *Species Journals*); Charles Darwin, *More Letters of Charles Darwin*, 2 vols. (New York: D. Appleton, 1903), 1:178 (hereafter cited as *MLD*); *Origin*, p. 65; Richard Owen, *Address* (London: n.p., 1858), p. 42; Hopkins, in Hull, *Darwin and His*

Critics, pp. 242–44, 268. Lyell, writing to Darwin in October 1859, contrasted natural selection favorably with the "purely unknown and imaginary" causes contained in "creation": *DLL,* 2:2.

10. A usage given philosophical sanction by Baden Powell, who called it "a mere *term of convenience*" adopted by "common consent." See his *Essays,* p. 399.

11. Charles Darwin, *Journal of Researches* (London: John Murray, 1897), p. 362 (hereafter cited as *Journal* [1845]); idem, *The foundations of the Origin of Species: Two Essays Written in 1842 and 1844,* ed. Francis Darwin (Cambridge: Cambridge University Press, 1909), pp. 155, 157–58 (hereafter cited as *Foundations*); idem, *Charles Darwin's Natural Selection,* ed. R. C. Stauffer (Cambridge: Cambridge University Press, 1975), pp. 575, 583–84 (hereafter cited as *Natural Selection*); idem, *The Collected Papers of Charles Darwin,* ed. Paul H. Barrett, 2 vols. (Chicago: University of Chicago Press, 1977), 2:92 (hereafter cited as *Collected Papers*). See also *DLL,* 1:394, 434.

12. [Robert Chambers], *Vestiges of the Natural History of Creation* (London: Churchill, 1844), pp. 127, 134, 144, 152, 184, 197. Chambers's ideas underwent changes over the years, but I shall deal only with the first edition of his book and his reply to his critics (1846). For a discussion of Chambers and his work in its entirety, see Milton Millhauser, *Just before Darwin: Robert Chambers and 'Vestiges'* (Middleton, Conn: Wesleyan University Press, 1959) and Bowler, *Fossils and Progress.*

13. *Origin,* p. 519. Owen was omitted from subsequent editions.

14. *American Journal of Science,* 2d ser., 1 (1846): 252; James Dwight Dana, "Science and the Bible," *Bibliotheca Sacra,* 13 (1856): 123–24; T. Vernon Wollaston, *On the Variation of Species* (London: Van Voorst, 1856), pp. 186–90; William Hopkins, "Anniversary Address of the President," *Journal of the Geological Society of London,* 8 (1852): lxxi; *The Times* (London), 26 December 1859, p. 8. The theoretical importance of creationism is suggested in Peter J. Bowler, "Evolution in the Enlightenment," *History of Science,* 12 (1974): 178–79.

15. Agassiz's influence was noted by, among others, ornithologist Alfred Newton, in "Early Days," p. 242, who found his views almost universal in America although much less so in England, and by General Joseph Portlock, who ranked him with H. G. Bronn as one of the two arbiters on the question of species. See "Anniversary Address of the President," *Journal of the Geological Society of London,* 14 (1858): clvi. The evidence suggests that miraculous creation was much more accepted among American naturalists than English. Why this should be so is not clear. The influence of Agassiz is probably part of the answer. The prominent evangelical protestantism of American men of science likely was a factor. At any rate, the question deserves study.

16. Louis Agassiz, *The Structure of Animal Life,* 2d ed. (New York: C. Scribner, 1866), pp. 92–111, 122; See also his comments on the discontinuity between fossil bearing formations in *AAAS Proceedings,* 3 (1850): 71, 73; idem, *Essay,* pp. 10–11, n. 8; 60, n. 77; 64–65, 90–91, 104. See his comment on J. W. Foster and J. D. Whitney, "On the Azoic System, as Developed in the Lake Superior Land District," *AAAS Proceedings,* 5 (1851): 7.

17. Edward Lurie, *Louis Agassiz, A Life in Science* (Chicago: University of Chicago Press, 1960), illustration following p. 84; Agassiz, *Essay,* pp. 44–45; Charles Lyell, *Life, Letters and Journals of Sir Charles Lyell, Bart,* ed. Mrs. Lyell, 2 vols. (London: John Murray, 1881), 2:331, 410 (hereafter cited as *LLL*); *Species Journals,* p. 265; Hooker, *Letters,* 1:474–75; Richard Owen, *On the Anatomy of Vertebrates,* 3 vols. (London: Longmans, Green, 1866–68), 3:805. For Darwin's low opinion of Agassiz as a scientific reasoner, see *MLD,* 1:104, 258, 309, 459, 476; 2:159, 161, 164, and Works Progress Administration, *Calendar of the Letters of Charles Robert Darwin to Asa Gray* (Boston: Historical Records Survey, 1939), pp. 56, 76 (hereafter cited as *Calendar*).

18. *MLD,* 1:262. Agassiz had considerable scientific insight within his frame of operation. Lyell's idea, for example, that each species was created as a single pair and then migrated over the earth made difficulties of ecological balance in an nonevolutionary system, as he himself unwittingly revealed elsewhere (*Principles* [1830], 2:124–25, 176–77). Agassiz's idea of the

creation of entire zoological provinces at once made better sense biologically, if in no other way. But for Lyell and others the incredibility of a miracle increased with its magnitude. It was Agassiz also who called for the study of specialized species with very narrow ranges, presumably on the theory that they were the most likely to be little modified by limited environmental adaption—a sort of controlled experiment—as a step toward a knowledge of the "conditions under which animals may have been created." See *Essay,* p. 38.

19. Richard Owen, "The Principal Forms of the Skeleton," in *Orr's Circle of the Sciences: Organic Nature* (London: Orr, 1854), 1:263; idem, *On the Classification and Geographical Distribution of the Mammalia . . . (London: John W. Parker, 1859),* appendix; idem, *Address* (1858), pp. 42–43; Sedgwick, *Discourse,* pp. cv, cxxxv–vi, ccxxxviii, 291. See also Sedgwick, in Hull, *Darwin and His Critics,* p. 161; John Phillips, *Life on Earth: Its Origin and Succession* (London: Macmillan, 1860), pp. 46, 213, 216; William Whewell, *History of the Inductive Sciences,* 3d ed., 3 vols. (London: John Parker, 1857), pp. 483–88; idem, review of the *Principles of Geology,* vol. 2, by Charles Lyell, *Quarterly Review,* 47 (1832): 117–18, 125; James D. Dana, "Agassiz's Contribution to the Natural History of the United States," *American Journal of Science,* 2d ser., 25 (1858): 208.

20. The Herschel letter is reprinted in full in W. F. Cannon, "The Impact of Uniformitarianism: Two Letters from John Herschel to Charles Lyell, 1836–37," *Proceedings of the American Philosophical Society,* 105 (1961): 301–14. See also *LLL,* 1:467–69. Both Cannon and Michael Ruse insist that Lyell envisioned a naturalistic, that is, positivistic, solution to the species problem. This would certainly make his continuing concern with theology in direct relation to speciation—as seen in his journals on species—anomalous. See Cannon, ibid., and "Lyell, Radical Actualism," pp. 111, 115, 202n.; Michael Ruse, "Charles Lyell and the Philosophers of Science," *British Journal for the History of Science,* 9 (1976): 128–29. It must be admitted, however, that Darwin used the same language as Herschel in 1845. But, since he was a transmutationist, the phrase manifestly had not the same meaning for him as it did for Herschel who was not. Once again, context must guide interpretation. See *DLL,* 1:394. Similarly, Lyell's deceptively "naturalistic" discussion of speciation in *The Geological Evidences of the Antiquity of Man,* 2d ed. (Philadelphia: Childs, 1863) (hereafter cited as *Antiquity of Man*), pp. 394–95, should be qualified by his other writings.

21. *DLL,* 1:550–51.

22. Whewell, *History,* 3:489; Duke of Argyll, *The Reign of Law,* 5th ed. (New York: John W. Lovell, n.d.), pp. 18, 134, 140–41, 156–60; *Species Journals,* p. 166. By 1863 Falconer had come to accept transmutation as "the most rational view" of species succession, but he still resisted Darwin's explanation of it (quoted in Gray, *Darwiniana,* pp. 159–60).

23. *MLD,* 1:418; Joseph Dalton Hooker, *Flora Novae Zealandicae,* in *The Botany of the Antarctic Voyage of H. M. Discovery Ships Erebus and Terror in the Years 1839–1843,* 3 vols. (London: Reeve, 1847–1860), 2:viii; idem, *Flora Tasmaniae,* in ibid., 3:ii, xxv–vi; idem, *Life and Letters of Sir Joseph Dalton Hooker,* ed. Leonard Huxley, 2 vols. (London: John Murray, 1918), 1:481–85. Although familiar with Darwin's views for fourteen years, Hooker did not adopt them until 1858. See his *Letters,* 1:447, 449, 520.

24. Thomas Henry Huxley, "Science and Religion," *The Builder,* 18 (1859): 35; *DLL,* 1:543; Wollaston, *Variation of Species,* p. 190; *Journals on Species,* pp. 56–57, 84, 165, 265–66, 284, 467–69; *LLL,* 2:212; Powell, *Essays,* p. 402; Huxley, *Letters,* 1:187.

25. *Species Journals,* pp. 355–56; Herbert Spencer, *An Autobiography,* 2 vols. (New York: D. Appleton, 1904), 1:176; idem, "The Development Hypothesis," *Essays, Scientific, Political and Speculative,* 3 vols. (New York: D. Appleton, 1896), 1:1–7. For Lyell's discussion see *Principles* (1830), 2:18–35. Thomas Henry Huxley, "On the Persistent Types of Animal Life," *Scientific Memoirs of Thomas Henry Huxley,* 4 vols. (London: Macmillan, 1898–1902), 2:92–93; Louis Agassiz, "Professor Agassiz on the Origin of Species," *American Journal of Science,* 2d ser., 30 (1860): 146.

26. *Species Journal,* pp. 200, 262 (brackets added by the editor); *Principles* (1830), 2:179–84; Hooker, *Flora Tasmaniae,* p. xxviii; Owen, *Anatomy,* 1:xxxvi. Also see Richard

Owen, *Palaeontology,* 2d ed. (Edinburgh: Black, 1861), pp. 441–44; Duke of Argyll, *Geology: Its Past and Present* (Glasgow: Griffin, 1859), pp. 29–30; Huxley, in *The Times* (London), 26 December 1859, p. 8; Asa Gray, *Natural Science and Religion* (New York: Scribner's, 1880), pp. 38–39; Gray, "Introductory Essay in Dr. Hooker's Flora of New Zealand: Vol. 1," *American Journal of Science,* 2d ser., 17 (1854): 340–48; *DLL,* 1:466; Powell, *Essays,* pp. 320, 397, 420–21; Phillips, *Life on Earth,* p. 205; Portlock, "Address" (1858), p. clvii; [Robert Chambers], *Vestiges of the Natural History of Creation with a Sequel* (New York: Colyer, 1846), pp. 203, 291 (hereafter cited as *Vestiges* [1846]).

Michael Bartholomew, "Lyell and Evolution: An Account of Lyell's Response to the Prospect of an Evolutionary Ancestry for Man," *British Journal for the History of Science,* 6, (1973): 261–303, argues convincingly that the reasons for Lyell's rejection of evolution were philosophical and theological rather than an extension of the logic of his uniformitarian geology, as others have held. For Owen's theoretical development, see Roy M. MacLeod, "Evolutionism and Richard Owen, 1830–1868: An Episode in Darwin's Century," *Isis,* 56 (1965): 259–80.

27. Rudwick, *Fossils,* p. 226.

28. John Herschel in *Victorian Science,* ed. George Basalla et al. (Garden City, N.Y.: Doubleday, 1970), pp. 406–7: Charles Bell, *The Hand* (London: William Pickering, 1834), pp. 43–44.

29. For Owen's idealism see his *On the Archetype and Homologies of the Vertebrate Skeleton* (London: Van Voorst, 1848), p. 172 ff. For Agassiz, see *Essay,* pp. 9–13, 17, 20–21, 23, 25–26, 43n., 51, 57, 72, 90–91, 98, 109, 130, 137.

30. James D. Dana, "Presidential Address," *AAAS Proceedings,* 9 (1854): 1–2; idem, "Thoughts on Species," *American Journal of Science,* 2d ser., 24 (1857): 307; John Phillips, "Anniversary Address of the President," *Journal of the Geological Society of London,* 16 (1860): xlviii–l; idem, *Life on Earth,* p. 25; Agassiz, "Origin of Species," pp. 146, 150–53; Gray, *Darwiniana,* p. 137; Owen, *Anatomy,* 3:795, 807–09; Powell, *Essays,* p. 400.

31. Sedgwick, *Letters,* 2:357–58.

32. E.g., *Principles* (1830), 1:85, 105, 164, 337; Hooker, *Flora Tasmaniae,* p. iv; Alfred Russel Wallace, "On the Law Which Has Regulated the Introduction of New Species" (1855), reprinted with notes in *Natural Selection and Tropical Nature,* rev. ed. (London: Macmillan, 1891), pp. 4–19; Wollaston, *Variation of Species,* pp. 7, 14, 72, 81; Andrew Murray, "On the Disguises of Nature," *Edinburgh New Philosophical Journal,* 68 (1860): 66–90.

33. Sedgwick, *Discourse,* pp. 212–22; William Whewell, *Astronomy and General Physics Considered with Reference to Natural Theology* (London: William Pickering, 1836), pp. 6–7; John Herschel, *A Preliminary Discourse on the Study of Natural Philosophy* (London: Longman, 1831), p. 100; John Stuart Mill, *A System of Logic,* 8th ed. (London: Longmans, 1961), pp. 206–8.

34. Gray, *Darwiniana,* p. 47; Herschel, *Discourse,* p. 37; Agassiz, *Essay,* pp. 127, 130–32; Owen, *Palaeontology,* p. 451; Chambers, *Vestiges,* pp. 25–26; Whewell, *Astronomy,* pp. 9–14; Sedgwick, *Discourse,* pp. clxxx, ccxxxviii, 98.

35. *Principles* (1830), 1:71–72; *DLL,* 1:450; Gray, *Letters,* 2:424.

36. Charles Darwin, *The Autobiography of Charles Darwin, 1809–1882,* ed. Nora Barlow (New York: Harcourt, Brace, 1958), p. 124 (hereafter cited as *Autobiography*); *DLL,* 2:13; Gray, *Darwiniana,* p. 164. Hooker, in 1853, found "very many naturalists" believing in permanent species, although "a large class" saw them as mutable; see Hooker, *Flora Novae Zealandicae,* p. ix; *Origin,* pp. 66, 750–51. For Owen's remonstrance, see *Anatomy,* 3:796, n. 2.

Chapter 3

1. *MLD,* 1:216. For explorations of Darwin's method in science, see Ghiselin, *Triumph;* A. C. Crombie, "Darwin's Scientific Method," *Actes du IXe congres, Academie Internationale d'Histoire des Sciences,* 1 (1960): 324–62; David L. Hull, "Metaphysics of Evolution," *British Journal*

for the History of Science, 3 (1967): 309–37; idem, *Darwin and His Critics;* Ernst Mayr, "Introduction" to *On the Origin of Species by Charles Darwin* (Cambridge: Harvard University Press, 1964); Rudwick, "Darwin and Glen Roy"; Michael Ruse, "Darwin's Debt to Philosophy: An Examination of the Influence of the Philosophical Ideas of John F. W. Herschel and William Whewell on the Development of Charles Darwin's Theory of Evolution," *Studies in History and Philosophy of Science,* 6 (1975): 159–81; idem, "Charles Darwin's Theory of Evolution: An Analysis," *Journal of the History of Biology,* 8 (1975): 219–42; Edward Manier, *The Young Darwin and His Cultural Circle* (Dordrecht, Holland: D. Reidel, 1978). Also see two reviews of Ghiselin's book: Frank N. Egerton, "Darwin's Method or Methods?" *Studies in History and Philosophy of Science,* 2 (1971): 281–86; Michael Ruse, "The Darwin Industry," *History of Science,* 12 (1974): 43–58.

2. Gruber, *Darwin on Man,* pp. 26, 46–72. On these and other possible influences, see Manier, *Young Darwin,* pt. 1.

3. *LLL,* 1:467.

4. *Autobiography,* pp. 69–70; Sedgwick, *Discourse,* pp. lii–iii, lxv, xxvi–vii, xlviii; *Letters,* 1:380. The most complete account of the trip (but containing none of my speculation) is Paul H. Barrett, "The Sedgwick-Darwin Geologic Tour of North Wales," *Proceedings of the American Philosophical Society,* 118 (1974): 146–64. Darwin's unpublished geological notes taken on the trip are included.

5. See, among numerous others, Gavin de Beer, *Charles Darwin: A Scientific Biography* (Garden City, N.Y.: Doubleday, 1964), p. 57; Rudwick, *Fossils,* pp. 188–89; Leonard G. Wilson, *Charles Lyell: The Years to 1841* (New Haven: Yale University Press, 1972), p. 447; Howard E. Gruber and V. Gruber, "The Eye of Reason: Darwin's Development during the Beagle Voyage," *Isis,* 52 (1962): 186–200. W. F. Cannon stresses Lyell's influence to the exclusion of all others, see "Lyell, Radical Actualism," pp. 117–18.

6. *LLL,* 1:234; *Autobiography,* p. 100; *MLD,* 2:117, Sedgwick, *Letters,* 2:412. On Darwin's independence of Lyell, see Ghiselin, *Triumph,* p. 15; Cannon, "Lyell, Radical Actualism," p. 118; and Michael Bartholomew, "The Non-progress of Non-progression: Two Responses to Lyell's Doctrine," *British Journal for the History of Science,* 9 (1976): 170–73.

7. *Autobiography,* p. 68; *MLD,* 1:263; Ruse, "Darwin's Debt." Martin Rudwick, who supports Ruse on this point, stresses Darwin's reliance on Herschel and Whewell for the "principle of exclusion" used by him in his abortive Glen Roy paper, but overlooks the fact that Paley and other natural theologians, who had been and were objects of Darwin's study, also used it to prove the existence of God from the order and adaptations of nature. Similarly, the "Herschelian phrase," *verae causae,* was taken by Herschel from Newton and was also common coin. Here, again, the general intellectual milieu could easily have been the source of the method. See Rudwick's impressive paper "Darwin and Glen Roy," pp. 169, 176. I do not dispute that Darwin was influenced by the scientists of his time on the question of how to do science; obviously, I insist on it. Nor do I doubt that Herschel was important, along with Lyell, Sedgwick, and the others. I simply question the implication that the works of Herschel and Whewell functioned as handbooks for Darwin, providing him with a cut-and-dried method. The evidence on the issue in both of these valuable papers is largely circumstantial—especially in Whewell's case—and much of the argument conjectural (as is mine on Sedgwick's influence). Rudwick's connecting of the structures of the Glen Roy paper and the essay of 1844 is striking and his point that Darwin may have felt that his failure on Glen Roy threatened his species theory is intriguing because the method of exclusion is not absent from his preselectionist work. In fact, the epistemic shift itself involved a type of exclusion. But this feature does not mean that both theories need have come out of the *Discourse on Natural Philosophy.* For Darwin's continued reading of natural theology through the 1840s see Peter J. Vorzimmer, "The Darwin Reading Notebooks (1838–1860)," *Journal of the History of Biology,* 10 (1977): 107–52. On the general importance of Darwin's early philosophical reading, see Manier, *Young Darwin.*

8. Charles Darwin, Transmutation notebook "E" (1838), p. 59, in Gruber, *Darwin on*

Man, p. 459; *DLL*, 2:37. For Herschel's later and more sympathetic assessment see Herschel, *Physical Geography* (Edinburgh: Black, 1861), p. 12n. On what it meant to be scientific see W. F. Cannon, "John Herschel and the Idea of Science," *Journal of the History of Ideas*, 20 (1961): 215–39; also Cannon, "Lyell, Radical Actualism"; Ruse, "Darwin's Debt."

9. Charles Darwin, *Darwin and Henslow: The Growth of an Idea, Letters 1831–1860*, ed. Nora Barlow (Berkeley: University of California Press, 1967), pp. 26, 55 (hereafter cited as *Darwin and Henslow*); *Diary*, pp. 20, 21, 24–25, 33, 39, 72–73; *MLD*, 1:263; *DLL*, 2:422. For Humboldt's importance, see Frank Egerton, "Humboldt, Darwin and Population," *Journal of the History of Biology*, 3 (1970): 325–60; and Cannon, "Lyell, Radical Actualism," pp. 118, 120, n. 26. On Darwin's love of nature, see John Angus Campbell, "Nature, Religion, and Emotional Response: A Reconsideration of Darwin's Affective Decline," *Victorian Studies*, 18 (1974): 159–74. Also see Francis Darwin's remarks in *DLL*, 1:93–95.

10. *Autobiography*, pp. 77–78, Ghiselin, *Triumph*, pp. 15–31; De Beer, *Darwin*, pp. 56–57; Crombie, "Scientific Method," pp. 358–59; Rudwick, "Darwin and Glen Roy," pp. 162–63; Gruber and Gruber, "Eye of Reason"; Charles Darwin, *The Structure and Distribution of Coral Reefs* (London: Smith, Elder, 1842). The theory is summarized in *Journal* (1839), pp. 555–69.

11. *Darwin and Henslow*, pp. 53, 65–66, 83, 102–7, 110; *Diary*, pp. 173–74, 261, 303–4, 322, 430; *Journal* (1839), pp. 157, 351, 373–81.

12. Gray, *Darwiniana*, pp. 142n. 144. Probably Gray was not atypical among Americans in his aversion to positivism. American scientists appear to have worked within the old episteme with striking harmony. For the peaceful relations of science and religion in America, see Stanley M. Guralnick, "Sources of Misconception on the Role of Science in the Nineteenth-Century American College," *Isis*, 65 (1974): 352–66.

13. Gruber, *Darwin on Man*, pp. 110–11.

14. *LLL*, 2:433–34.

15. The degree to which religious orthodoxy did forcefully frustrate geological research has been quite properly questioned. See, for example, Rhoda Rapaport, "Geology and Orthodoxy: The Case of Noah's Flood in Eighteenth-Century Thought," *British Journal for the History of Science*, 11 (1978): 1–18. But the danger offered by biblicism, as Lyell realized, lay not in repression in the name of biblical literalism or some prescribed orthodox interpretation of Genesis, but in the flood motif and in other persistent conventions of thought among geologists.

16. *Principles* (1830), 1:5–11, 29–30, 76, 104–5, 148–49, 302, 442; 3:148–49; *LLL*, 1:173–74, 268–71, 328. For Lyell's resentment at the role of the clergy in promoting bad geology, see *LLL*, 2:168, 172, For Lyell's place as a scientific polemicist, see Roy Porter, "Charles Lyell and the Principles of the History of Geology," *British Journal for the History of Science*, 9 (1976): 91–103; W. F. Cannon, "Charles Lyell is Permitted to Speak for Himself: An Abstract," in Cecil J. Schneer, ed., *Toward a History of Geology*, (Cambridge: MIT Press, 1969), pp. 78–79; and Martin J. S. Rudwick, "The Strategy of Lyell's *Principles of Geology*," *Isis*, 61 (1970), 5–33.

17. James D. Dana, "On American Geological History," *American Journal of Science*, 2d ser., 22 (1856): 331; idem, *Manual of Geology*, 3d ed. (New York: Ivison, 1880), pp. 845–50; Sedgwick, *Discourse*, pp. 111–12, 115; *Letters*, 2:76–79; *The Times* (London), 30 September 1844, p. 6; Owen, *Address* (1858), pp. 3, 34–36; Charles Lyell, "Anniversary Address of the President," *Journal of the Geological Society of London*, 6 (1850): xxxiii; Stallo, quoted in Numbers, *Creation by Natural Law*, p. 74.

18. Hooker, *Flora Novae Zealandicae*, pp. viii–xii, xxvi; idem, *Flora Tasmaniae*, p. xiii; Gray, "Introductory Essay," p. 243–44; idem, *Darwiniana*, pp. 196–97; *Principles* (1872), 2:332–33; *Origin*, p. 568; Wollaston, *Variation of Species*, pp. 4–5.

19. Huxley, "Science and Religion," p. 35; idem, *Science and the Christian Tradition* (New York: D. Appleton, 1898), pp. vii–viii; Hooker, *Letters*, 1:477–48, 516, 520; Agassiz, *Essay*, p. 12n.

20. Charles Lyell, "Review of the Transactions of the Geological Society of London, 1824," *Quarterly Review,* 34 (1826): 538–40; *Species Journals,* pp. 405–7.

21. *LLL,* 1:316–18. For Lyell's moderate religious views and his generally temperate relations with the clergy, see Bartholomew, "Lyell and Evolution," pp. 267–68, 298–99.

22. T. Neville George, "Charles Lyell, The Present Is Key to the Past," *Philosophical Journal: Transactions of the Royal Philosophical Society of Glasgow,* 13 (1976): 11; Wilson, *Lyell,* pp. 309–13.

23. *LLL,* 2:337.

24. *Principles* (1830), 3:270–74, 384–85. Bishop Colenso, who employed the volcanic cones of the Auvergne as one of his arguments against the flood's historicity (a point Lyell himself had made in his letter to the bishop of Llandaff: see Wilson, *Lyell,* p. 310), pointed out that there is nothing in the biblical account to suggest that the waters were miraculous or "preternatural" in origin: see John William Colenso, *The Pentateuch and Book of Joshua Critically Examined,* 2d ed. (London: Longmans, Green, 1862), p. viii. Lyell, of course, was indignant at the charge that he had been guilty of mystery-mongering at any point in the *Principles: LLL,* 1:467. Perhaps this solution, by no means original with Lyell, occurred to him after his correspondence with the bishop of Llandaff and so was included in vol. 3. For the theory of a nonviolent Mosaic flood, see Leroy E. Page, "Diluvialism and Its Critics in Great Britain in the Early Nineteenth Century," in Schneer, *History of Geology,* pp. 257–71. Page does not indicate whether advocates of this theory considered the flood to have been "preternatural" in origin. See also Gordon L. Davies, *The Earth in Decay* (New York: Science History Publications, 1969), pp. 37–38.

25. Argyll, *Geology,* pp. 48–49; *LLL,* 2:375–76. See *Species Journals,* pp. 118, 121, 233, 330–31, 336 for continuing theological elements in Lyell's science.

26. *DLL,* 1:507; 2:24, 34, 48, 69, 168, 315; *MLD,* 1:202, 236, 272, 333, 308–9; 2:232. See also Gruber, *Darwin on Man,* p. 325.

27. Agassiz, "Origin of Species," pp. 148–49; idem, *Structure of Animal Life,* pp. vi, 1–2, 128; Lurie *Agassiz,* p. 353; *MLD,* 2:160; *DLL,* 2:63; Gray, *Darwiniana,* pp. 16–18.

28. Foucault, *Order,* pp. 158–59. For the close relation of idealism and an emphasis on taxonomy, see Mary P. Winsor, *Starfish, Jellyfish and the Order of Life* (New Haven: Yale University Press, 1976).

29. Agassiz considered biology in the mid-nineteenth century to be a completed science in its major dimensions and that evolutionists had added nothing important to its body of knowledge. See his *A Journey in Brazil* (Boston: Houghton Mifflin, 1895), p. 7; Hull, *Darwin and His Critics,* p. 436. Lawrence Badash, "The Completeness of Nineteenth-Century Science," *Isis,* 93 (1972): 48–58, discusses this attitude in physics. Possibly this represents the maturation of positivism in that science while biology was just entering the same phase. Agassiz's position would then signify the maturation of the old episteme in biology.

30. *DLL,* 1:266; Anon., "Review of *Cours de philosophie positive* by Auguste Comte," *Edinburgh Review,* 67 (1838): 145–62. For a similar scheme, see *Principles* (1830), 1:76. Darwin, notebook "N" (1838), p. 12, in Gruber, *Darwin on Man,* p. 332; notebook "M" (1838), pp. 69–70, 135–36, in ibid., pp. 278, 291–92. Darwin also cited "M. LeCompte" in support of his argument against free will, notebook "M," pp. 72–74, in ibid., pp. 278–79. For his dislike of the later cult of Positivism, see *DLL,* 2:328 and *MLD,* 1:313, 382. The best discussions of Darwin and Comte are in Silvan S. Schweber, "The Origin of the *Origin* Revisited," *Journal of the History of Biology,* 10 (1977): 231, 241–64; and Manier, *Young Darwin,* pp. 40–47 and passim, esp. p. 167. See also Sandra Herbert, "The Place of Man in the Development of Darwin's Theory of Transmutation, Part II," *Journal of the History of Biology,* 10 (1977): 220, 224.

31. *MLD.* 1:192, 194. See also Charles Darwin, *The Descent of Man and Selection in Relation to Sex,* 2d ed. (New York: D. Appleton, 1897), p. 24, n. 56 (hereafter cited as *Descent*).

32. Sedgwick had a similar understanding of natural law (see *Discourse,* pp. 98, 212–22), as did Herschel and Whewell (see Ruse, "Darwin's Debt," pp. 159–81).

33. *Origin*, p. 165. For similar operational definitions of "species" and "varieties," see p. 136.

34. For examples of law as metaphor, see *Origin*, pp. 86, 118, 317, 435, 471, 632, 644, 688, 729, 738, 758; *Descent*, pp. 5, 8, 460, 497, 548, 554, 614. As generalizations see *Origin*, pp. 158, 191, 378–79, 565, 711, 736, 742, 744; *Descent*, pp. 6, 29, 30, 48, 146, 190, 222, 227, 232, 271, 360, 466, 500, 585. For metaphysical usage, see *Origin*, pp. 423, 757–58. For his rejection of a "fixed law of development," see *Origin*, pp. 523, 567.

35. Charles Darwin, *The Variation of Animals and Plants under Domestication*, 2d ed., 2 vols. (New York: D. Appleton, 1896), 1:9 (hereafter cited as *Variation*). Also see *Origin*, p. 342; *Calendar*, pp. 20, 56; *DLL*, 2:37, 80. Lyell gives a similar definition of theory in *Antiquity of Man*, p. 395. On the meaning of "explain," see *DLL*, 2:58, Charles Darwin, notebook no. 2 (1838), p. 55, in "Darwin's Notebooks on Transmutation of Species," ed. by Gavin De Beer, *Bulletin of the British Museum (Natural History), Historical Series*, 2, no. 3 (1960): 87 (hereafter cited as *Bulletin*); notebook no. 1 (1836–37), pp. 226–29, *Bulletin*, 2, no. 2 (1960): 69; *Origin*, pp. 185, 720. For this same view in Lyell, see Rudwick, *Fossils*, p. 170.

36. *Origin*, pp. 111, 131; *Descent*, p. 423, n. 35; 434, 525, 613–14. Lyell had a similar understanding of chance as ignorance of law (*Principles* [1830], 2:51), as did Huxley (*DLL*, 1:553). For Darwin's usage in the conventional theological sense, see *Autobiography*, pp. 92–93 and *MLD*, 1:321, 395. Darwin's idea of chance in relation to natural selection was similar to Mill's view of natural events "casually" rather than "causally" connected: *Logic*, pp. 345–46.

37. *Principles* (1830), 1:72, 131, 160, 165; also 88, 92–93, 150, 431; Lyell, "Anniversary Address of the President," *Journal of the Geological Society of London*, 7 (1851): lxxiv. Wilson attributes Lyell's analogical thinking to the influence of Bishop Butler (*Lyell*, p. 281), but quoting Butler in early nineteenth-century England was as conventional as quoting Shakespeare. Surely a specific influence is not needed to account for an idea so general in science.

38. *Foundations*, pp. 102–3; *Origin*, pp. 752–53; *MLD*, 2:67; *Collected Papers*, 1:215; Transmutation notebook no. 4 (1838–39), p. 128, in *Bulletin*, 2, no. 5 (1960): 174.

39. Mill, *Logic*, p. 368; David B. Wilson, "Herschel and Whewell's Version of Newtonianism," *Journal of the History of Ideas*, 35 (1974): 93–94. For skepticism about Darwin's domestic analogy, see the famous review by Fleeming Jenkin in Hull, *Darwin and His Critics*, pp. 305–12.

40. *MLD*, 2:80; *DLL*, 1:439; Robert M. Stecher, "The Darwin-Inness Letters: The Correspondence of an Evolutionist with His Vicar, 1848–1884," *Annals of Science*, 17 (1961): 225–26. For Darwin's varied use of analogy, see Ghiselin, *Triumph*, and two papers by Michael Ruse, "The Nature of Scientific Models: Formal v. Material Models," *Philosophy of the Social Sciences*, 3 (1973): 63–80 and "The Value of Analogical Models in Science," *Dialogue*, 12 (1973): 246–53. In his "Darwin's Debt," pp. 175–76, Ruse sees a formal mode of proof in Darwin's use of analogy. Unfortunately, his argument is almost entirely circumstantial and is based on the assumption that Darwin was carefully following Herschel's prescriptions for scientific method. A close study of Darwin's writings will, I think, show a more varied and looser use of analogy. I doubt that Darwin ever attempted to *prove* anything by analogy alone, certainly not natural selection.

41. *Natural Selection*, p. 223; *Origin*, pp. 568, 717; *Principles* (1830), 1:39; 3:2–3. On the heuristic value of theories see *DLL*, 2:147, 204, 207, 210; *MLD*, 2:65. See Kavaloski, "*Vera Causa* Principle," on the importance of *verae causae* in Victorian science, especially in relation to the problems posed by special creation.

42. Emil Du Bois-Reymond, "The Limits of Our Knowledge of Nature," *Popular Science Monthly*, May 1874, p. 23; Cannon, "Lyell, Radical Actualism," pp. 107, 113–18; R. Hooykaas, *The Principle of Uniformity in Geology, Biology and Theology* (Leiden: Brill, 1963), pp. 4–47; Rudwick, *Fossils*, pp. 110n, 119, 185–87; idem, "Uniformity and Progression: Reflections on the Structure of Geological Theory in the Age of Lyell" in Roller, *Perspectives*.

43. Roderick I. Murchison et al., *The Geology of Russia in Europe and the Ural Mountains*, 2

vols. (London: John Murray, 1845), 1:532–33, 545–46; William Whewell, "On the Wave of Translation in Connexion with the Northern Drift," *Journal of the Geological Society of London,* 3 (1847): 231–32.

44. Lyell, "Presidential Address" (1850), p. xlv.

45. Rudwick, *Fossils,* p. 179; Cannon, "Lyell, Radical Actualism," pp. 110–11, *Principles* (1830), 1:61, 112; 2:159; 3:6–7, 148–49.

46. *LLL,* 1:234; 2:3–7.

47. Ghiselin, *Triumph,* p. 14; Ruse, "Lyell," p. 126; Cannon, "Lyell, Radical Actualism," p. 114; Leroy E. Page, "The Rivalry between Charles Lyell and Roderick Murchison," *British Journal for the History of Science,* 9 (1976): 158; Kavaloski, "*Vera Causa* Principle," pp. 87–91.

48. *LLL,* 1:269–70, 328; *Principles* (1830), 1:4, 11, 156–57; 3:383–85; *Species Journals,* p. 183; Wilson, *Lyell,* pp. 277, 285.

49. *Principles* (1830), 1:155–64; Powell, *Essays,* pp. 438–39.

50. Darwin, printed in Peter J. Vorzimmer, "An Early Darwin Manuscript: The 'Outline and Draft of 1839,'" *Journal of the History of Biology,* 8 (1975): 216.

51. Sedgwick, *Discourse,* p. cxxxviii; *Origin,* pp. 263, 265, 526–27, 566; Natural Selection, pp. 346, 350, 358, 511.

52. Gruber, *Darwin on Man,* pp. 62–63; Kuhn, *Scientific Revolutions,* p. 84; *MLD,* 1:189; *Calendar,* pp. 81, 85; Charles Darwin, "Some Unpublished Letters of Charles Darwin," *Royal Society of London Notes and Records,* 14 (1959): 52 (hereafter cited as "Unpublished Letters"). For his distrust of mathematicians as scientific reasoners see *MLD,* 1:153–54, 314, 336. For Mill's more extensive comments see Mill, *Logic,* pp. 327–28n. On the induction problem, see Hull, *Darwin and His Critics,* pp. 6–36. Hull's contention that Darwin misunderstood Mill's support of his position is puzzling since the point of that support was not that Darwin had proved his theory, which, Mill rightly noted, Darwin did not claim, but that, in Mill's opinion, Darwin had not been guilty of unscientific procedure. This seems clearly to have been Darwin's understanding and Mill's meaning. See Hull, ibid., pp. 7–8, 27–28.

53. *Variation,* 1:14. Similar statements fill his writings, e.g., *Foundations,* pp. 193–94; *Origin,* p. 544; notebook no. 3 (1838), pp. 71, 117, in *Bulletin,* 2, no. 4 (1960): 138, 142; *Collected Papers,* 2:79, 84.

54. *MLD,* 1:455. For Darwin's comments on ad hoc theorizing, see ibid. 1:56, 172, 335, 412, 461, 484; 2:15, 21–22, 135; *DLL,* 1:431; 2:127. Michael Ghiselin has laid to rest the prevalent charge that Darwin himself indulged in carefree ad hoc speculation; see *Triumph,* passim. Martin Rudwick traces Darwin's aversion to ad hoc explanations to the influence of Lyell and Herschel: see "Darwin and Glen Roy," p. 157. Sometimes accused of using natural selection in an ad hoc manner, Darwin actually condemned such practice: see *Origin,* pp. 364–65; *MLD,* 1:198; 2:63.

In alluding to the ad hoc nature of pangenesis, I do not intend to subscribe to the tradition that it was invented to fend off Fleeming Jenkin and Lord Kelvin and was not a long-considered and serious feature of Darwin's thought. Against this tradition, see Gerald L. Geison, "Darwin and Heredity: the Evolution of His Hypothesis of Pangenesis," *Journal of the History of Medicine,* 24 (1964): 375–411. It would seem that Darwin could, on occasion, use ad hoc theories quite creatively: see E. David Kohn, "Charles Darwin's Path to Natural Selection" (Ph.D. diss., University of Massachusetts, 1975), for a discussion of his preselectionist theories.

55. *DLL,* 2:228, 255, 260–61. 262; *MLD,* 1:301; 2:71, 371; *Autobiography,* p. 130. He could, however, sound a more confident note: *DLL,* 2:256, 264–65, 301; *MLD,* 2:427; *Collected Papers,* 2:160. See also his defense of pangenesis against Galton's experiments with rabbit's blood: *Collected Papers,* 2:166. On Lamarck, see Gruber, *Darwin on Man,* p. 193. Also see *Principles* (1830), 2:8.

56. Notebook no. 1 (1836–37), p. 104, in *Bulletin,* 2, no. 2 (1960): 53. Also see *DLL,* 2:13.

57. *MLD,* 1:139–40, 173; *DLL,* 1:437, 485; 2:10, 36–37, 482; notebook "B" (1837),

pp. 227–29, in Gruber, *Darwin on Man,* p. 447; *Origin,* pp. 754–57; *Principles* (1830), 1:460–62; John Tyndall, "Scientific Materialism," in *Fragments of Science,* 2 vols. (New York: D. Appleton, 1897), 2:77.

58. *MLD,* 1:126; *DLL,* 2:145, 147, 207, 398. Also see *Calendar,* p. 90; *Collected Papers,* 2:81; and "Unpublished Letters," p. 35.

59. *DLL,* 1:117, 122, 125; 2:103, 237, 392; *MLD,* 1:176, 195; 2:133, 324; "Geology" in John F. W. Herschel, ed., *A Manual of Scientific Enquiry,* 2d ed. (London: John Murray, 1851), pp. 173–74. This essay is also in *Collected Papers,* 1:227–49. See also *Autobiography,* p. 141. He was sometimes reproached with arguing too well against his own views—see *MLD,* 1:269; *DLL,* 2:252—and sometimes amused himself by composing mental essays against his theories: *DLL,* 2:143. The passage on Spencer (*Autobiography,* p. 109) is omitted from the version printed in *DLL.* See also *DLL,* 2:371; *MLD,* 2:235, 424–25, 442. On observation see *Autobiography,* p. 159; Emma Darwin, *Emma Darwin: A Century of Family Letters, 1796–1896,* ed. Henrietta Litchfield, 2 vols. (New York: D. Appleton, 1915), 2:207. Lyell had a similar view: see *Principles* (1830), 1:459.

60. *MLD,* 1:150, 184, 323. The reference is to John Tyndall's lecture, "On the Scientific Uses of the Imagination," *Fragments,* 2:101–34. In connection with Darwin's suggested parallel with physics, see Geoffrey Cantor, "The Reception of the Wave Theory of Light in Britain: A Case Study Illustrating the Role of Methodology in Scientific Debate," *Historical Studies in the Physical Sciences,* 6 (1975): 109–32; F. W. Hutton, "Some Remarks on Mr. Darwin's Theory," *Geologist,* 4 (1861): 188; *DLL,* 2:155; Hooker, *Flora Tasmaniae,* p. iv *Antiquity of Man,* pp. 471–72.

61. *Descent,* p. 61; *Origin,* pp. 749–50; *Calendar,* p. 27. Also see *Collected Papers,* 2:81. The quotation from Whewell was: "But with regard to the material world, we can at least go so far as this—we can perceive that events are brought about not by insulated interpositions of Divine power, exerted in each particular case, but by the establishment of general laws." That from Bacon was: "To conclude, therefore, let no man out of a weak conceit of sobriety, or an ill-applied moderation, think or maintain, that a man can search too far or be too well studied in the book of God's word; or in the book of God's works; divinity or philosophy; but rather let men endeavour an endless progress of proficience in both."

Chapter 4

1. *Diary,* p. 383; *Journal* (1839), pp. 53, 399–400, 400n, 526–27. In the 1845 edition of the *Journal* two of the creationist passages were deleted: *Journal* (1845), pp. 44, 313.

2. *DLL,* 2:15, 211; *MLD,* 1:234, 367. In *The Expression of the Emotions in Man and Animals* (New York: D. Appleton, 1898) (hereafter cited as *Expression*), which appeared in 1872, Darwin speaks of himself as having been "inclined to believe" in evolution in 1838 (p. 18). Francis Darwin was convinced that his father was fully converted to evolution when he kept the notebooks of 1837–38; *MLD,* 1:38. Most writers have agreed with him. On the problem of the date of his adoption of transmutation see Camille Limoges, *La sélection naturelle* (Paris: Presses universitaires de France, 1970); Sandra Herbert, "The Place of Man in Development of Darwin's Theory of Transmutation, Part 1 to July, 1837," *Journal of the History of Biology,* 7 (1974): 216–58. On Darwin's being "tempted" to revert to special creation—however seriously—as late as early 1838, see Gruber, *Darwin on Man,* p. 153.

3. *Foundations,* pp. 22–23. For the same passage in the 1844 essay, see pp. 133–34.

4. Ibid., pp. 248, 253–54. Also see notebook "B" (1837), cited in Gruber, *Darwin on Man,* p. 111.

5. *Origin,* p. 749. Darwin detested idealist verbalisms in science: *MLD,* 1:73; *Descent,* p. 24. Richard Owen, *On the Nature of Limbs* (London: Van Voorst, 1849), pp. 68–69; Sedgwick, *Discourse,* p. xii; Darwin, cited in Gruber, *Darwin on Man,* pp. 417–18; also see pp. 196–98.

Foundations, pp. 38–40, 47, 202, n. 3; also see *Origin,* p. 712. Chambers made much of Macleay's system as evidence of intelligent design in nature in his early editions; see *Vestiges,* pp. 238–51.

6. *Origin,* pp. 649, 675–76, 683.

7. *MLD,* 1:202. On this feature of the work, see Ghiselin, *Triumph,* chap. 6.

8. Charles Darwin, *On the Various Contrivances by Which British and Foreign Orchids Are Fertilised by Insects and on the Good Effects of Inter-crossing* (London: John Murray, 1862), pp. 1–2, 285, 287–88, 306–7 (hereafter cited as *Orchids*).

9. Miller, *Footsteps,* pp. 283–84.

10. Notebook no. 1 (1836–37), p. 141, in *Bulletin,* 2, no. 2 (1960): 58. Darwin did not reject a functionally conceived teleology, but only one that implied a predetermining purpose. He acknowledged that, in the former sense, his own work was teleological in that adaptive function was at its heart. See *DLL,* 2:367; also Gray, *Darwiniana,* pp. 237, 293–94.

One of the often remarked paradoxes of Darwin's writing was his continuing use of purposeful teleological language, as well as other telic metaphors, to express functional teleological concepts. This practice is analogous to his use of the language of animal behavior to describe the behavior of plants. Habit, the conventions of his time, and the interesting but neglected problem of developing a suitable scientific langauge account for this practice. Confusing as it is—and was—I think those are mistaken who have read philosophical problems and contradictions into it. This can only be done by an obviously misplaced literalism. St. George Jackson Mivart was somewhat unusual among Darwin's theologically grounded critics in recognizing the metaphorical nature of this usage and dropping it as an object of criticism: see *On the Genesis of Species* (London: Macmillan, 1871), pp. 14–15. More typical was Argyll, *Reign,* pp. 22–24, 153–54.

11. *MLD,* 1:144–45. Beauty in the world was a problem that Darwin never solved. Its functional uses under natural or sexual selection he could reveal, but why certain colors and structures were perceived as beautiful by man and, presumably, by other higher animals was a puzzle he could not resolve. The origin of beauty stood in the same relation to sexual selection as the origin of variations did to natural selection: a continuing bafflement and, because of its traditional associations with natural theology, a potential threat. See *Origin,* pp. 367, 369–72, 737; *Descent,* pp. 92–93, 261–62, 495, 569, 584; *DLL,* 2:248.

12. *Origin,* pp. 113, 372–73; *Natural Selection,* pp. 382–83, 466; *DLL,* 2:67; Whewell, *Astronomy,* pp. 4–5, 20–29.

13. *Origin,* pp. 185, 375.

14. *DLL,* 2:133; *Foundations,* p. 182; *Origin,* pp. 308–9. 361; *Expression,* pp. 19, also 11–12; *Variation,* 1:9.

15. *Origin,* pp. 300, 304, 739; *Natural Selection,* pp. 336–38, also 307, 318, 333; *Descent,* pp. 39, 43.

16. *Origin,* pp. 120, 139, 143, 273, 312, 424, 440–41, 470, 471, 739–40, 755; also see *MLD,* 1:248; *Foundations,* pp. 50, 96, 204; *Natural Selection,* p. 146; *Descent,* p. 176. Darwin also noted the tendency of creationists such as Agassiz to multiply the number of species needlessly: *DLL,* 1:439.

17. *Origin,* pp. 553, 703–5. On homologies, rather than the mere recapitulation of adult progenitors, being the key to Darwin's interest in embryology, see Stephen Jay Gould, *Ontogeny and Phylogeny* (Cambridge: Harvard University Press, 1977), pp. 70–73. Also see Dov Ospovat, "The Influence of Karl Ernst von Baer's Embryology, 1828–1859: A Reappraisal in Light of Richard Owen's and William B. Carpenter's 'Palaeontological Law,'" *Journal of the History of Biology,* 10 (1976): 1–28.

18. *Origin,* pp. 568, 642–43; *Foundations,* pp. 156, 162, 168–73, 182, 246; for 1842 sketch, see pp. 31, 33–34. Transmutation notebook no. 2 (1838), pp. 160, 184, 205, 240 and no. 3 (1838), pp. 54–56, in *Bulletin,* 3, no. 5 (1967): 136, 150–51, 155, 159. Lyell suggested a creationist alternative to caprice in the island-continent problem: "that species[,]

if created to struggle with American forms, would have to be created on American type." Darwin rejected the underlying supposition of consistency of type as contrary to fact: see *DLL*, 2:5.

19. *Foundations*, p. 216; *Origin*, pp. 676–83.

20. *Origin*, pp. 705–16; *Natural Selection*, p. 296; *Foundations*, pp. 231–33.

21. *Origin*, pp. 459–60, 487–509, 524, 527; *Foundations*, pp. 136–38. See also Hooker, *Flora Tasmaniae*, pp. xxii–xxiv.

22. This was the logic that underlay Cuvier's classic objection to transmutation: see Owen, *Anatomy*, 3:789.

23. Agassiz, Hopkins, and Jenkin, in Hull, *Darwin and His Critics*, pp. 442–44; 265, 269; and 342–43, respectively.

24. Agassiz, *Essay*, pp. 28, 103; idem, "The Primitive Diversity and Number of Animals in Geological Times," *American Journal of Sciences*, 2d ser., 17 (1854): 309–24; Sedgwick, *Discourse*, pp. xxvii, lix, lxxii, lxxvii, xcviii, cxxiii; *Letters*, 2:361.

25. *MLD*, 1:163, 173.

26. *Foundations*, p. 107.

27. *Variation*, 1:9; *Origin*, pp. 627, 677–78; *Natural Selection*, p. 374; *Orchids*, pp. 348–49, 351.

28. *Origin*, pp. 209, 284–85, 502; *Natural Selection*, pp. 388–90, 409. See also Charles Lyell's presidential address to the British Association for the Advancement of Science, *The Times* (London), 15 September 1864, p. 8.

29. Notebook no. 3 (1838), p. 115, in *Bulletin*, 2, no. 4 (1960): 141; *Foundations*, p. 52, n. 1.

30. *Natural Selection*, p. 286. The quoted passage was lined out by Darwin in the manuscript.

31. *Origin*, pp. 591, 619, 623–25, 631, 640, 744–45.

32. Ibid., pp. 335–36; *Natural Selection*, p. 347.

33. *Journal* (1839), pp. 60, 106–8, 111–12, 144–45; *Journal* (1845), p. 49; *Natural Selection*, pp. 349, 380, 527.

34. *Natural Selection*, p. 167. For a related creationist paradox produced by mimicry, see *Collected Papers*, 2:90.

35. *Origin*, pp. 278, 470, 734; *Natural Selection*, pp. 280, 331–32.

36. *Origin*, pp. 140, 344, 367, 372, 414, 422, 476, 526; *Natural Selection*, pp. 390, 566.

37. *Variation*, 2: 186; *DLL*, 2:10, 182; *MLD*, 1:198, 209, 234, 304, 312; 2:379. Manier comments on the fuzziness of Darwin's "tests" in *Young Darwin*, pp. 11–12, 36.

Chapter 5

1. *Origin*, pp. 750–51. Darwin later conceded that this assault had been "too rude," but his defensive remarks do not alter the point at issue: see *DLL*, 2:106.

2. *Origin*, p. 267.

3. *Natural Selection*, p. 565; *Origin*, p. 572; *Descent*, pp. 61, 181, 608; *DLL*, 2:134–36; *MLD*, 1:377; Ruse, "Two Biological Revolutions," p. 19.

4. *MLD*, 1:161; *DLL*, 2:110. On the difficulties of natural selection as a metaphor, see *DLL*, 2:111, 139, 230; *MLD*, 1:126–27, 153, 175, 271, 369; Asa Gray, *Letters*, 2:626–27.

5. William H. Harvey, "Darwin on the Origin of Species," *Gardener's Chronicle and Agricultural Gazette*, 18 February 1860, pp. 145–46; *DLL*, 2:69–70. The reference to "manufacture" applies to Darwin's use of that metaphor as an alternative to creation: see *Origin*, p. 140.

6. Relevant passages are William Paley, "Natural Theology" in *Works of William Paley, D. D.* (London: William Tegg, 1853), pp. 26–27, 43, 72, 85–89, 104–6, 107–10, 110–12, 115–17, 119–24, 128, 166–67; Peter Roget, *Animal and Vegetable Physiology*, 2 vols. (London: W. Pickering, 1834), 1:23–24, 29–30, 32–34; Bell, *The Hand*, pp. xi, 169,

244–45; William Buckland, *Geology and Mineralogy Considered with Reference to Natural Theology,* 2 vols. (London: W. Pickering, 1836), 1:44–46, 49, 58–59, 76, 114, 164, 186, 233, 332, 356, 370, 381, 403, 523, 547, 584, 588–89; William Kirby, *On the Power, Wisdom and Goodness of God as Manifested in the Creation of Animals and in Their History, Habits and Instincts,* 2 vols. (London: W. Pickering, 1835), 1:138–39, 199, 225, 294, 323, 359; 2:91, 98, 109–10, 173, 220, 281, 394; William Prout, *Chemistry, Meteorology and the Function of Digestion Considered with Reference to Natural Theology* (London: W. Pickering, 1834), pp. 3–7, 14, 23. Also see Sedgwick, *Discourse,* pp. x–xiii.

7. Owen, *Limbs,* pp. 9–15, 39–40, 84–86; Agassiz, *Essay,* pp. 11–12, 71–72; Powell, *Essays,* p. 367. Peter Bowler has reached similar conclusions regarding the importance of idealist design: see "Darwinism and the Argument from Design," and *Fossils and Progress,* p. 45. For a similar and earlier shift in the design argument in cosmology, see Numbers, *Creation by Natural Law,* p. 80.

8. Duke of Argyll, "Mr. Herbert Spencer and Lord Salisbury on Evolution," *Nineteenth Century,* April 1897, p. 570; Mivart, *Genesis,* p. 270.

9. *MLD,* 1:203.

10. *Autobiography,* p. 59; *DLL,* 2:15; Transmutation notebook no. 3 (1838), p. 49, in *Bulletin,* 2, no. 4 (1960): 134. On Darwin's ironic debt to Paley, see Cannon, "Darwin's Achievement."

11. *Foundations,* pp. 15–16, 128–32; *Origin,* pp. 337–43.

12. Transmutation notebook no. 3 (1838), p. 74, in *Bulletin,* 3, no. 5 (1967): 160; *MLD,* 1:372, 395; *Descent,* p. 613.

13. *Calendar,* pp. 27–28, 40; *DLL,* 2:146, 169–70.

14. For an example of polarization, see his letter to Gray in *Calendar,* p. 94; also *DLL,* 1:285n. This meeting seems to have occurred in February 1881: see Emma Darwin, *Letters,* 2:245.

15. Chambers, *Vestiges,* pp. 152–59, 164, 184–85, 188–89, 205–11; *Vestiges* (1846), p. 269.

16. Chambers, *Vestiges,* pp. 73, 87, 89–90, 146–49, 176, 203–4.

17. Ibid., pp. 157–58, 192, 194, 213–17, 219, 222–23, 228–30, 232–33.

18. Ibid., pp. 152–65, 168, 176, 184–85, 197–98, 223, 232–33, 237, 324; *Vestiges* (1846), p. 202.

19. Owen, *Anatomy,* 3:799–801.

20. Ibid., pp. 788–89, 797–98, 808–9.

21. Ibid., 1:v–vi; ix, x–xi; 3:790–91, 796, 808. Also *Archetype,* pp. 171–72.

22. Owen, *Anatomy,* 3:795, 797, 807–9, 817–18.

23. For Argyll's relation to Owen, see Duke of Argyll, *Autobiography and Memoirs,* ed. Dowager Duchess of Argyll, 2 vols. (London: John Murray, 1906), 1:408–11, 581.

24. Duke of Argyll, *The Unity of Nature* (New York: G. P. Putnam's Sons, 1885), pp. 34–35, 44, 298–304; idem, *The Philosophy of Belief, or, Law in Christian Theology* (London: John Murray, 1896), p. 116.

25. Argyll, *Unity,* pp. 275–76, 289–93, 304–5; idem, *Reign,* pp. 120–23.

26. Duke of Argyll, "A Fourth State of Matter," *Nature,* 22 (1880): 168; *Philosophy of Belief,* p. 28.

27. Duke of Argyll, "The Power of Loose Analogies," *Nineteenth Century,* December 1887, p. 751.

28. Argyll, *Philosophy of Belief,* p. 74; idem, *Unity,* pp. 287–88, 292–93.

29. Herbert Spencer, *The Principles of Sociology,* 5 vols. (New York: D. Appleton, 1896), vol. 1, pt. 1, p. 361.

30. See, for instance, Argyll, *Reign,* pp. 19–20, 26–27, 66; idem, *Autobiography,* 1:328–31, 405–7.

31. Argyll, *Philosophy of Belief,* pp. xv, 4–5, 181; idem, *Unity,* p. 44; idem, *Reign,* pp. 19–20, 236.

32. Argyll, *Philosophy of Belief,* pp. 18–19, 50–51.

33. Argyll, *Unity,* pp. 145–51, 207, 278–79; idem, *Reign,* p. 66.

34. Argyll, *Unity,* p. 165.

35. Ibid., pp. 306–7.

36. Argyll, *Philosophy of Belief,* pp. 8–9.

37. Ibid., pp. 23, 185–86, 199.

38. Edward Burnett Tylor, *Primitive Culture,* 2 vols. (London: John Murray, 1871), 2:86; J. Foulerton, "The Duke of Argyll on the Unity of Nature," *Journal of Science,* 3d. ser., 3 (1880): 672–73. For Argyll's critique of evolutionary anthropology, see Neal C. Gillespie, "The Duke of Argyll, Evolutionary Anthropology and the Art of Scientific Controversy," *Isis,* 68 (1977): 40–54.

39. Argyll, *Reign,* pp. 20–21, 76–77, 104, 152; idem, "Spencer and Salisbury," p. 571.

40. Ibid., pp. 391–92; Hooker, *Letters,* 2:124–25; Argyll, *Nature,* 25 (1881–82): 7. In this exchange with Romanes, one of Argyll's many brief scientific duels, he continued the paradoxical Baconian tradition of ostensibly avoiding mixing science and theology while advocating a science that was steeped in theological presuppositions; see *Nature,* 24 (1881): 505–6, 581, 604; 25 (1881–82): 7, 29–30.

41. Argyll, *Reign,* p. 136.

42. Ibid., pp. 10–14, 18–19, 30–33, 129, 140.

43. Ibid., pp. 17–18, 156; idem, "Loose Analogies," pp. 748–49, 755–57; idem, "Spencer and Salisbury," pp. 396, 570.

44. Duke of Argyll, *Primeval Man: An Examination of Some Recent Speculations* (New York: J. W. Lovell, n.d.), pp. 14–15, 23. These essays were first published in the periodical *Good Words* in 1868. Idem, *Reign,* pp. 127–29, 133–34, 140, 156–60; idem, *Philosophy of Belief,* p. 72.

45. Argyll, *Reign,* pp. 17–18, 134, 140–41, 155–59.

46. Argyll, "Spencer and Salisbury," pp. 388–90, 395, 400, 402, 572–73, 584. Argyll's late essays on evolution were published as *Organic Evolution Cross-examined or, Some Suggestions on the Great Secret of Biology* (London: John Murray, 1898). His ambivalence on evolution is also apparent in his *What is Science?* (Edinburgh: David Douglas, 1898), pp. 30–63.

47. Argyll, *Unity,* p. 268; idem, *Philosophy of Belief,* pp. 172–76; idem, "Spencer and Salisbury," pp. 570, 574.

48. Argyll, *Unity,* pp. 262–73; idem, "Spencer and Salisbury," pp. 400–401, 575; idem, *Reign,* pp. 122–23; idem, *Nature,* 38 (1888): 341–42, 564, 615.

49. Argyll, *Reign,* pp. 140–43, 158–59; idem, *Primeval Man,* p. 15; idem, "Spencer and Salisbury," pp. 397–99, 571, 573. See also *LLL,* 2:413.

50. Argyll, "Spencer and Salisbury," pp. 391, 399, 401, 578–79; idem, "Loose Analogies," pp. 746–48; idem, *Reign,* pp. 22, 48–49, 112–14, 117–18, 131, 133, 137–40, 146; idem, *Unity,* p. 291; idem, "On Mr. Wallace's Theory of Birds' Nests," *Journal of Travel and Natural History,* 1 (1868–69): 286–87.

51. Typical of these encounters, and the most serious, was that involving the Duke's charge of a "reign of terror" against non-Darwinists and the suppression of data unfavorable to Darwin's coral reef theory. See Argyll, "A Great Lesson," *Nineteenth Century,* September 1887, pp. 293–309; T. H. Huxley, "Science and the Bishops," ibid., November 1887, pp. 625–41; and then *Nature,* 37 (1887–88): 25–26, 53–54, 246, 272, 293, 317–18, 342, 363–64. Also see Huxley, *Letters,* 2:168–72, 288; idem, *Science and Christian Tradition,* pp. 417–19; idem, *Science and Hebrew Tradition* (New York: D. Appleton, 1896), pp. 283–86. For the last of these clashes, which shows the growing professional impatience with Argyll, see *Nature,* 59 (1898–99): 217–19, 246–47, 269, 317–18.

52. [W. J. Youmans], "The Duke of Argyll on Evolution," *Popular Science Monthly,* July 1897, p. 411. Also see Anon., review of the *Philosophy of Belief* by the Duke of Argyll, *Athenaeum,* 13 June 1896, pp. 771–72; and Argyll's obituary notice: Anon., "The Duke of Argyll as a Naturalist," *Athenaeum,* 28 April 1900, p. 532. Of all Darwin's group, Hooker was perhaps the most scathing: see Hooker, *Letters,* 2:114.

53. Mivart's effectiveness as a critic of Darwin, a subject I shall not take up, is a matter of dispute. Peter Vorzimmer's depiction of Mivart "badgering" Darwin "into a state of frustrating confusion" seems overdrawn: *Charles Darwin: The Years of Controversy, 1859–1882* (Philadelphia: Temple University Press, 1970), pp. 225–51. On the other side, see Michael T. Ghiselin, "Mr. Darwin's Critics, Old and New," *Journal of the History of Biology*, 6 (1973): 155–65; and Darwin's reply to Mivart in *Origin*, pp. 242–67, 353–54. Also see Hull, *Darwin and His Critics*, pp. 412–15; Jacob W. Gruber, *A Conscience in Conflict: The Life of St. George Jackson Mivart* (New York: Columbia University Press, 1960), pp. 52–69, 84–114. On Darwin's ire, see *DLL*, 2:315, 323–29; *MLD*, 1:332–34.

54. Mivart, *Genesis*, pp. 125, 237–38, 259, 270, 275–76.

55. Ibid., pp. 16–18, 252–53, 258–59, 262, 277, 288.

56. Ibid., pp. 4, 15, 156, 186–87, 225.

57. Ibid., pp. 254–58, 270.

58. Ibid., pp. 47–48, 63, 96–97, 102, 111–12, 118, 121–22, 127, 276, 288.

59. Ibid., pp. 103, 133–34, 143, 229.

60. Ibid., pp. 233–36, 238–40.

61. Ibid., pp. 5, 19–21, 63, 66–69, 72–78, 231, 240. In a real sense, Mivart's critique was seriously misdirected since it was intended throughout as an attack on the proposition that evolution was the result of the action of natural selection alone, a position Darwin never took. See ibid., pp. 19–20, and passim.

62. Ibid., pp. 3–5; also chaps. 9 and 12.

63. Chambers, *Vestiges*, p. 189.

Chapter 6

1. *Principles* (1830), 1:69; 2:41–42, 44; Charles Lyell, *Principles of Geology*, 8th ed. (London: John Murray, 1850), pp. 572–73, 575 (hereafter cited as *Principles* [1850]); *Species Journals*, pp. 168, 176–77, 413, 418–23, 424, 449–50, 458–59, 497; *DLL*, 2:9–10, 384–85; *Principles* (1873), 2:495–96.

2. *Antiquity of Man*, pp. 408, 469, 505–6; *LLL*, 2:442; *Principles* (1872), 2:318, 496–97, 499–500. In his refusal to accept the hypothetical sequences that Darwin suggested for the evolution of the eye, Lyell made a common error by misrepresenting Darwin's argument. Darwin's intention was not to show how the eye had evolved but to suggest that it could have done so within the confines of his theory. He gave many such hypothetical explanations, but did not confuse them with reality; see, e.g., *Natural Selection*, pp. 341, 346, 350, 358, 511.

3. Gray, *Letters*, 1:321, 345–46; 2:476, 480. For a full discussion of Gray's relations with Darwin see A. Hunter Dupree, *Asa Gray, 1810–1888* (New York: Atheneum, 1968).

4. Asa Gray, Review of *On the Various Contrivances by which . . . Orchids are Fertilised . . .* by Charles Darwin, *American Journal of Science*, 2d ser., 34 (1862): 139; idem, *Letters*, 2:485–86, 488–89, 508, 512; *MLD*, 1:242.

5. Charles Hodge, *What is Darwinism?* (New York: Scribner, 1874). Gray, *Darwiniana*, pp. xi–xii, 43–48, 51–71, 117–26, 139, 142, 193, 213–14, 223–25, 229, 248, 311–12, 315–19; idem, *Science and Religion*, pp. 62–66, 85–91. Dupree's contention that as late as 1860 a "designed" view of natural selection was a real option for Darwin seems doubtful in the light of the evidence presented herein. See *Asa Gray*, pp. 300–301, 339.

6. Gray, Darwiniana, pp. 16–18, 41–42, 62, 121–22, 129, 130, 161, 193.

7. George J. Romanes, "Natural Selection and Natural Theology," *Contemporary Review*, 42 (1882): 536–39, 543.

8. George J. Romanes, "Natural Selection and Natural Theology," *Nature*, 27 (1882–83): 362–64, 528–29. Asa Gray, "Natural Selection and Natural Theology," *Nature*, 27 (1882–83): 291–92, 527–28; idem, *Letters*, 2:742–43; idem, *Science and Religion*, pp. 94–95.

9. Alfred Russel Wallace, "The Limits of Natural Selection as Applied to Man [1870]," reprinted with notes in *Natural Selection and Tropical Nature*, new ed. (London: Macmillan, 1891), pp. 187–88, 204–5, 205n–6n; idem, *On Miracles and Modern Spiritualism: Three Essays*,

2d ed. (London: Trübner, 1881), pp. vi–viii, 31, 36–37, 40–43, 46, 100–101, 106–8, 124–25; *DLL,* 2:297. On Wallace's spiritualism and natural selection, see Malcolm Jay Kottler, "Alfred Russel Wallace, the Origin of Man, and Spiritualism," *Isis,* 65 (1974): 145–92.

10. Gray's "shot" was: "But in Mr. Darwin's parallel [between the architect and the process of natural selection], to meet the case in nature according to his own view of it, not only the fragments of rock (answering to variation) should fall, but the edifice (answering to natural selection) should rise, irrespective of will and choice!"—*The Nation,* 19 March 1868, p. 236. See also *Variation,* 2:248–49; Gray, *Letters,* 2:562, 613, 627.

11. Joseph LeConte, *Evolution and Its Relation to Religious Thought,* 2d ed. (New York: D. Appleton, 1891), pp. 270, 346; Carpenter, *Nature and Man,* pp. 409–63; Hooker, *Letters,* 1:518–19; 2:106; Hutton, in Hull, *Darwin and His Critics,* p. 299; Dana, *Geology,* pp. 603–4; W. Stanley Jevons, "Evolution and the Doctrine of Design," *Popular Science Monthly,* May 1874, pp. 98–100; F. C. S. Schiller, "Darwinism and Design," *Contemporary Review,* 71 (1897): 879–83; Huxley, *Letters,* 1:490; 2:200–201; *DLL,* 1:555.

12. *Origin,* pp. 73, 264–65, 751; *DLL,* 2:84; *MLD,* 1: 244, 367, 384. On saltation and on species as "discrete entities," see David L. Hull, *Philosophy of Biological Science* (Englewood Cliffs, N.J.: Prentice-Hall, 1974), pp. 52–53.

13. Huxley, *Lay Sermons,* p. 297. One of the most curious aspects of the relation between Huxley and Darwin was Huxley's inability to grasp the nuances of natural selection and associated doctrines: see *MLD,* 1:139–40, 225, 230–31, 274; 2:233. After Darwin's death, Huxley complained to Hooker of the difficulty of the *Origin:* Huxley, *Letters,* 2:204. On Huxley's shortcomings as a Darwinian, see Michael Bartholomew, "Huxley's Defence of Darwin," *Annals of Science,* 32 (1975): 525–35.

14. Most large mutations seemed to occur in domestic races. Because of the abnormal conditions under which they lived, Darwin thought their variability was also abnormal; that is, while not differing in kind from that of natural populations, it did differ in degree and frequency. Domestic saltations, therefore, could not be treated as replicas of natural variations. Domestic variation demonstrated the relative plasticity of the living form and provided a laboratory for the study of heredity; but, while analogically suggestive, it was not an adequate testing ground for true speciation or the process of variation leading to it. By this line of reasoning, Darwin could justify downplaying the importance of domestic saltations; but he did not *know* that such jumps did not occur in nature. All he could say was that they did not seem to be significant. See *Variation,* 1:2–5; *Origin,* pp. 77–83, 120–22, 145; *Natural Selection,* pp. 106, 323–24; *The Effects of Cross and Self Fertilisation in the Vegetable Kingdom* (New York: D. Appleton, 1895), p. 287 (hereafter cited as *Cross Fertilisation*).

15. *MLD,* 1:166, 198; 2:6–7, 163–64, 379; *DLL,* 2:126–27, 197, 288, 296. *Origin,* pp. 264–65, 405, 412, 513–14; *Natural Selection,* p. 319. On Jenkin, see Peter Vorzimmer, "Charles Darwin and Blending Inheritance," *Isis,* 54 (1963): 371–90; Hull, *Darwin and His Critics,* pp. 344–50; Peter J. Bowler, "Darwin's Concepts of Variation," *Journal of the History of Medicine,* 29 (1974): 196–212. For Kelvin, see Joe D. Burchfield, *Lord Kelvin and the Age of the Earth* (New York: Science History Publications, 1975), pp. 70–86; Mivart, *Genesis,* pp. 136–42. For a discussion of Darwin's problem with saltation as a biological fact and as a source of theoretical difficulty, see Vorzimmer, *Charles Darwin,* pp. 46–69. Vorzimmer notes that it was the discontinuous nature of saltation that primarily worried Darwin.

16. A function discussed by Robert M. Young, "Darwin's Metaphor: Does Nature Select?" *Monist,* 55 (1971): 477–85. For the antecedents of designed evolution, see Bowler, "Evolution in the Enlightenment," pp. 176–77.

17. *DLL,* 2:267, 297; *LLL,* 2:442–43.

18. *DLL,* 2:6–7, 28; *MLD,* 1:190–92. It is interesting that Darwin was anxious to spread Gray's views on design and selection in England, even though he personally rejected them. Conversion to natural selection on any grounds was better than none. See *MLD,* 1:169–70; *DLL,* 2:163; also, Dupree, *Asa Gray,* pp. 298–300.

19. *Origin,* pp. 77–80, 164, 166, 178–80, 234, 275–80, 692; *Natural Selection,* pp. 279–80; *Descent,* pp. 28, 30, 44, 177–78, 608; *Variation,* 2:237, 240–41, 255, 260–61, 280–82; *DLL,* 2:516–17; *Cross Fertilisation,* pp. 310–11, 446–47.

20. Gruber, *Darwin on Man,* pp. 105–6; *Descent,* p. 62.

21. *DLL,* 2:97–98, 139, 146, 170, 192, 321; *MLD,* 1:192, 321; *Calendar,* p. 52; "Unpublished Letters," p. 35.

22. *DLL,* 1:283–84; 2:165–66; *MLD,* 1:194.

23. *DLL,* 2:228; *MLD,* 2:371; also see *MLD,* 1: 301; 2:82; *Variation,* 2:349–99.

24. *Autobiography,* p. 130; *MLD,* 1:335; 2:427; *Variation,* 2: chaps. 24–26; *Descent,* pp. 232–42.

Chapter 7

1. Dana, "Agassiz's Contributions," p. 202.

2. Gruber, *Darwin on Man,* p. 58.

3. *DLL,* 1:132. Kuhn remarks on such shifting vision as characteristic of a period of "paradigm crisis." See *Scientific Revolutions,* p. 114.

4. *Principles* (1830), 2:133; *Species Journals,* pp. 358, 427–28; also pp. 413, 415, 459. Gray's response to the moral problem of cruelty in nature was similar: see his *Darwiniana,* pp. 128–29, 306–11. Mivart, like some theologians, resolved the problem by simply denying that the physical suffering of brutes—mental suffering was out of the question—was a moral evil, and that beasts, like the lower races of men, felt much pain anyway. See Mivart, *Genesis,* pp. 260–61.

5. For example, Mivart, *Genesis,* pp. 259–60.

6. *DLL,* 2:105–6; *MLD,* 1:94; *Origin,* pp. 373–75, 392, 394, 737–38; Gray, *Darwiniana,* p. 130. Darwin himself later provided a rationale for the waste of pollen: "the production of a great and apparently wasteful amount of pollen [is] necessary for cross-fertilisation." Yet, if credited with rationalizing an evil, rather than with explaining an adaptive mechanism, he might have asked why an omnipotent God would not have devised a more efficient way to insure the genetic health of plants: see *Cross Fertilisation,* p. 3.

7. *Foundations,* pp. 51–52. For the same passage in the 1844 essay see p. 254. Words erased by Darwin have been omitted. On the problem of reconciling animal as well as human suffering with the existence of God, see *DLL,* 1:276, 284; 2:377; *MLD,* 1:395; *Calendar,* pp. 27–28, 31. Also Donald Fleming, "Charles Darwin, the Anaesthetic Man," *Victorian Studies,* 4 (1961): 230–34. Fleming mistakes this theological difficulty for a rejection of religion.

8. *Natural Selection,* pp. 381–82, 386. Chambers employed a similar argument for God's lack of responsibility. While the laws governing inorganic nature are fairly simple and no exceptions occur in their operation, he said, those ruling organic nature (and the weather) are so complex and contradictory in operation that a slipshod performance is all we can reasonably expect even from God. But if the details escape divine control, the result is good in general. Chambers held out the hope to men that there was more to the divine plan than met the eye, but, as with Mivart, what happened to other animals did not seem to concern him. Chambers, *Vestiges,* pp. 363–86.

9. Paley, "Natural Theology," pp. 168–71.

10. For Darwin's ambivalence regarding "higher" and "lower" as well as "progress," see *Origin,* pp. 220–21, 241, 547–54, 688, 757–58; *MLD,* 1:164; 2:76; *DLL,* 1:384; 2:6, 56, 89; *Collected Papers,* 2:79; *Descent,* pp. 140, 145, 164, 618–19. See also Bowler, *Fossils,* chap. 6.

11. *Origin,* pp. 162, 423; also see pp. 168–69; *Natural Selection,* pp. 208, 527; *Diary,* pp. 212–13.

12. Sedgwick, *Discourse,* pp. lxxiv–lxxxvi, cxxxv–vi, ccxliii; *DLL,* 2:44. Darwin's ambivalence regarding struggle in nature was paralleled by his ambivalence about it among men: see *Descent,* pp. 133–34, 142–43, 618. Also see John C. Greene, "Darwin as a Social

Evolutionist," *Journal of the History of Biology,* 10 (1977): 1–28.

13. Transmutation notebook no. 1 (1836–37), pp. 193–94, 218, in *Bulletin,* 2, no. 2 (1960): 64–65, 67; *Foundations,* p. 49, in 1844 essay, pp. 249–51; *Origin,* pp. 317, 751; *Descent,* pp. 25, 146. Also see *MLD,* 1:145.

Philip Gosse's *Omphalos,* for all its eccentricity, was a disturbing assault on the unexamined assumptions underlying the common belief of most naturalists in the use of analogy and in the uniformity of nature. Gosse, though converting few, seems to have made many aware, as Darwin was aware, of the need for things to be what they seem if science was to be possible. General Portlock, for instance, spent three pages of his presidential address before the Geological Society of London denouncing Gosse and defending the metaphysics of positive science. See Portlock, "Anniversary Address" (1858), pp. clix–clxi. For a similar sensitivity to this type of skepticism, see Argyll, *Reign,* p. 159.

14. *Foundations,* pp. 87, 182, 251; *Origin,* pp. 757–58; *MLD,* 1:154; Transmutation notebook no. 1 (1836–37), p. 98, in *Bulletin,* 2, no. 2 (1960): 53; notebook "M," p. 154e, in Gruber, *Darwin on Man,* p. 296; Darwin quoted in Herbert, "Place of Man," p. 233.

15. Transmutation notebook no. 1 (1836–37), p. 216, in *Bulletin,* 2, no. 2 (1960): 67. Michael Bartholomew, "Non-progress of Non-progression," p. 170, has suggested, without pressing the point, that this could be a slap at Lyell's nonprogressive view of organic nature. Perhaps—the ambiguity is sufficient. Yet, to me, it makes better sense as a condemnation of the idea of the necessary special creation of each new species, however closely allied to its predecessors.

16. Notebook "N" (1838), p. 36, in Gruber, *Darwin on Man,* p. 337. Additions by Darwin have been omitted. Transmutation notebook no. 1 (1836–37), pp. 101–2, 114, in *Bulletin,* 2, no. 2 (1960): 53, 55; *DLL,* 2: 247.

17. See Spencer, "Development Hypothesis."

18. In the *Descent,* Darwin said that one of his goals in writing the *Origin* was to discredit special creation: see p. 61. In a letter to Gray in 1863, he described "Creation *or* Modification" as the main issue, with natural selection only secondary: *DLL,* 2:163–64; also pp. 198, 207; *MLD,* 1:304.

19. He had little patience with metaphysical puzzles, especially if connected with science: *MLD,* 1:192–94.

20. Ibid., 1:273; 2:171; *Variation,* 1:12.

21. *Natural Selection,* pp. 225, 581; *Origin,* p. 751. Also see *Foundations,* p. 37; *MLD,* 1:173; *DLL,* 2:6.

22. Transmutation notebook no. 2 (1838), p. 102, in *Bulletin,* 3, no. 5 (1967): 147; *Athenaeum,* 28 March 1863, p. 419; 25 April 1863, pp. 554–55 (also in *Collected Papers,* 2:78–79); "Unpublished Letters," p. 59; *DLL,* 2:346–47; *MLD,* 1:163–64, 273, 321–22; 2:171; *Origin,* p. 223. On the question of spontaneous generation, see Farley, *Spontaneous Generation,* pp. 121–42.

23. *Origin,* pp. 343–44; *DLL,* 2:245. In the original version of this passage on the eye, Darwin concluded, "May we not believe that a living optical instrument might be found, as much superior to one of glass, as the works of Nature are to those of Art?" He subsequently struck out "Nature" and "Art" and wrote in "the Creator" and "Man." This can, to be sure, be seen as metaphor or "truckling," but why make the change at all? Perhaps, given his anxiety over the question of the eye, giving God even an indirect role in its formation—creation by natural selection (and wildly contradictory of his science as this was)—reassured him. See *Natural Selection,* pp. 352–53.

24. *Origin,* p. 748; *DLL,* 2:32, 82; *Foundations,* p. 254; Transmutation notebook no. 3 (1838), pp. 36–37, in *Bulletin,* 2, no. 4 (1960): 132. Darwin's aversion to an anthropomorphic view of the Creator is noted by George Grinnell, "The Rise and Fall of Darwin's First Theory of Transmutation," *Journal of the History of Biology,* 7 (1974): 262–63.

25. *Variation,* 2:431. See also *Calendar,* p. 74; Gray, *Darwiniana,* pp. 113, 122.

Chapter 8

1. *DLL,* 2:202–3. Lady Nora Barlow, in her remarks on Darwin's ornithological notes, extends the period of regret back beyond the *Origin* to the earliest evolutionary writings. It is not clear why "long regretted" could not apply to a duration of four years. Her interpretation involves Darwin in incredibly obscure behavior: "Darwin's Ornithological Notes," ed. by Nora Barlow, *Bulletin,* 2, no. 7 (1963): 207. Howard Gruber explains Darwin's use of the term creation as owing to a habit "in harmony with his abiding desire to placate his religious critics," and suggests that in the early notebooks the usage "may have reflected his own religious doubts." Gruber *Darwin on Man,* p. 154, n. 2; also see pp. 209–12. Both Barlow and Gruber employ the Hooker letter to dismiss all of Darwin's "creation" passages. See also Farley, *Spontaneous Generation,* p. 80, and Hull, *Darwin and His Critics,* pp. 52–53 for other acceptances of this as truckling.

2. *Calendar,* p. 46.

3. *Autobiography,* pp. 92–93; *DLL,* 2:412.

4. *DLL,* 2:45.

5. The "breathed" image occurs also in *Origin,* p. 759, where the phrase "by the Creator" was added in the second edition, as well as in the 1842 sketch (*Foundations,* p. 52) and in *Natural Selection,* p. 248.

6. Huxley seemed to share Darwin's apprehension, censoring Darwin for "taking refuge in 'Pentateuchal phraseology' ": Huxley, *Letters,* 1:261–63. Darwin, for his part, acknowledged that the usage was "perhaps" inappropriate in a scientific work: *Collected Papers,* 2:78–79.

7. Edward Aveling, *The Religious Views of Charles Darwin* (London: Free Thought Publishing Company, 1883), p. 7.

8. For example, Herbert, "Place of Man," pp. 219–20; Gruber, *Darwin on Man,* pp. 43, 75, 207; Ghiselin, "The Individual," p. 122. In the second part of her careful analysis of the Darwin notebooks, Herbert modifies her position somewhat, acknowledging that "traditional religion" or Christianity was what Darwin gave up. She continues, however, to neglect the significance of Darwin's theological sentiments. See "Place of Man, II," pp. 201–2; 202, n. 85; 211. Silvan Schweber has given us one of the most thorough attempts to date to reconstruct Darwin's "loss of religion" from the transmutation notebooks. In my judgment, he fails to show that Darwin was "agnostic (and possibly an atheist)" in 1839, largely because he treats religion and Christianity as synonymous, assumes much too quickly that materialism was atheism, and seems to equate a loss of belief in divine intervention in nature with a loss of religion. See Schweber, "The Origin of the *Origin,*" pp. 233–34, 304–10. I should add that the point at issue is not the main burden of Schweber's valuable essay. Manier, in *Young Darwin,* is more cautious, but nonetheless discounts Darwin's theism for similar, if not identical, reasons: see pp. 68, 87–89. Both of these accounts, however, are necessary reading on the question. Maurice Mandelbaum carries Darwin's theism through the period of the *Origin,* but has him a thorough agnostic from the 1860s on: see "Darwin's Religious Views."

9. *Autobiography,* pp. 85–87.

10. Herbert, "Place of Man," p. 223; *Autobiography,* pp. 56–57.

11. *Diary,* pp. 8, 14, 17, 75, 110, 243–44; *MLD,* 2:31. Ussher's chronology was, of course, printed in the King James version of the Bible for many years.

12. *Autobiography,* p. 59; *Journal* (1839), pp. 29, 604–5.

13. *Autobiography,* pp. 236–39; Emma Darwin, *Letters,* 2:173–75. Their daughter Henrietta mentions her mother's Christian devotion and misgivings about her husband's work: ibid., 2:173, 196.

14. *Autobiography,* p. 100. Lyell's writings, especially his species journals, show that this apparently straightforward declaration of his liberalism should not be misconstrued as theological indifference. It was probably the anti-Mosaic and antibiblicist aspects of Lyell's thought that most impressed young Darwin.

15. Ibid., p. 87. This passage indicates that the existence of a personal deity did not

become a *problem* for Darwin until later in his life. During the period prior to the 1860s it probably was held as a residue of his Christianity and provoked little or no defensive speculation.

16. *DLL,* 2:357; Aveling, *Religious Views of Darwin,* pp. 5–6. Francis Darwin, who was present, endorses Aveling's account somewhat ambiguously, saying that Aveling "gives quite fairly his impression of my father's views." He does not, however, question the direct quotations attributed to Darwin: *DLL,* 1:286n.

In 1879 Darwin replied to a query about the compatibility of evolution and religion: "Science has nothing to do with Christ, except in so far as the habit of scientific research makes a man cautious in admitting evidence. For myself, I do not believe that there ever has been any revelation." See *DLL,* 1:277.

17. Howard Gruber, who hesitates to say that Darwin was either a materialist or a non-theist but who seems to believe that he was both from an early period, opened this important question. See *Darwin on Man,* especially pp. 104, 209; Gruber's uncertainty is shared by Manier, *Young Darwin,* pp. 86, 167–68, 195–96. On this point also see Schweber, "The Origin of the *Origin.*"

18. Notebook "C" (1838), p. 166, in Gruber, *Darwin on Man,* pp. 450–51.

19. Notebook "M," p. 57, ibid., p. 276.

20. Quoted in Gruber, ibid., p. 201. Gruber acknowledges that Darwin's limitations on materialism were not incompatible with theism; see p. 104.

21. Notebook "M," pp. 19, 27, 31, 72–74, 101e in Gruber, ibid., pp. 269, 270–71, 278–9, 284; notebook "N," p. 5, ibid., p. 331.

22. Ibid., p. 215; also Schweber, "The Origin of the *Origin,*" p. 310. Manier's discussion of the materialism issue is perceptive; see *Young Darwin,* pp. 56–68, 129–34, 159–63.

23. Notebook "M," p. 74 in Gruber, *Darwin on Man,* p. 279; also pp. 180–81; Sedgwick, *Letters,* 2:360; idem, *Discourse,* pp. xi–xix, xlix–l, cxl–clvi. The importance and novelty of Darwin's making no exception for man in his positivism is stressed by Herbert, "Place of Man, II," pp. 196–207.

24. Tyndall, *Fragments of Science,* 2:78–79, 85–89, 107–8, 128–34, 168–69, 201; Huxley, *Lay Sermons,* pp. 144–46; Karl von Nägeli, "The Limits of Natural Knowledge," *Nature,* 4 (1871): 563; *Expression,* pp. 352–54; *MLD,* 1:398–99. Huxley, of course, denied being a materialist: *Methods and Results* (New York: D. Appleton, 1898), p. 245.

25. See the Duke of Argyll's *The Reign of Law* (1866) which was written to reassure Christians on this point. The marriage of positivism and faith was the foundation of many, if not of most, schemes of accommodation. Owen also defended "materialism" in science; see *Anatomy,* 3:821–25.

26. *Autobiography,* p. 57.

27. Ibid., pp. 87–94; *DLL,* 1:274.

28. *DLL,* 2:146; *MLD,* 1:321, 395; 2:171, 237; "Unpublished Letters," p. 35. Pusey's sermon in the *Guardian,* 20 November 1878, p. 1612, said of Darwin, "It [the theological doctrine of the First Cause] is foreign to the researches of physical science, so much so, that when one who traced the development of species through all links possible and impossible, closed his book, not as a philosopher but *as an atheist,* by speaking of life and its several powers having been originally breathed by the Creator into a few forms or into one, the expression was criticized, because it acknowledged God as the Author of all" (italics added). The printed sermon reads "as a theist." Indeed, the apparent typographical error in the *Guardian* article makes no sense. See E. B. Pusey, *Un-Science, Not Science, Adverse to Faith,* 2d ed. (London: Parker, 1878), p. 54; *DLL,* 2:412. For Francis Darwin, see *DLL,* 1:286n.

29. *DLL,* 1:284–86. Here, again, Darwin used "chance" in the conventional Victorian sense of purposeless activity unguided by intelligence.

30. The text of the letter is in Gruber, *Darwin on Man,* p. 27, n. 13. On the question of the addressee of the letter, whether Marx or Edward Aveling, see Lewis S. Feuer, "Is the Darwin-Marx Correspondence Authentic?" *Annals of Science,* 32 (1975): 1–12; Lewis S. Feuer

et al., "On the Darwin-Marx Correspondence," ibid., 33 (1976): 383–94; Margaret A. Fay, "Did Marx Offer to Dedicate *Capital* to Darwin?" *Journal of the History of Ideas,* 39 (1978): 133–46.

31. For example, Chadwick, *Secularization,* p. 167.

32. *DLL,* 2:34, 82, 345, 412; *MLD,* 1:309; Alan Brown, *The Metaphysical Society: Victorian Minds in Conflict, 1869–1880* (New York: Columbia University Press, 1947), p. 18; Stecher, *Letters,* pp. 244–45. Pusey's doctrine of separation is in *Un-Science,* pp. 5–7.

33. *DLL,* 2:24, 34, 48, 69, 315; *MLD,* 1:202, 272, 308–9, 333; 2:232; *Origin,* p. 40.

34. For in intensive inquiry into the complex problem of the association of Darwin's evolutionary thought and his illness, and a review of earlier theories, see Ralph Colp, Jr., *To Be an Invalid: The Illness of Charles Darwin* (Chicago: University of Chicago Press, 1977).

Epilogue

1. Kuhn, *Scientific Revolutions,* p. 76.

2. Hull, *Darwin and His Critics,* p. 386.

3. Ibid., p. 139; *Species Journals,* p. 241; *DLL,* 1:541, 548; Tyndall, *Fragments of Science,* 2:29. On the general issue of miracles and science, see Alvar Ellegård, *Darwin and the General Reader* (Göteborg: University of Gothenburg, 1958), chap. 7, and W. F. Cannon, "The Problem of Miracles in the 1830's," *Victorian Studies,* 4 (1960): 5–32. For the United States, see John F. McElligott, "Before Darwin: Religion and Science as Presented in American Magazines, 1830–1860" (Ph.D. diss., New York University, 1973), pp. 60–70.

4. *Species Journals,* e.g., 328, 408–9, 415, 421, 426–27.

5. Anon., "The Immutability of Nature," *Quarterly Review,* 110 (1861): 368–400; Henry Goodwin, Bishop of Carlisle, "The Uniformity of Nature," *Popular Science Monthly,* December 1885, p. 255.

6. Tyndall, *Fragments of Science,* 2:12, 32–34; also 2:1–6; Sedgwick, *Discourse,* pp. xliii, cxxxv; Huxley, "Scientific and Pseudo-Scientific Realism," *Nineteenth Century,* February 1887, pp. 191–205; Duke of Argyll, "Professor Huxley on Canon Liddon," ibid., March 1887, pp. 321–39; Huxley, "Science and Pseudo-Science," ibid., April 1887, pp. 481–98; Argyll, "Science, Falsely so-called," ibid., May 1887, pp. 771–74.

7. Huxley, *Science and Christian Tradition,* pp. 79–83.

8. Ibid., pp. xiii, 198; Argyll, "Huxley on Liddon," p. 322; Carpenter, *Nature and Man,* pp. 241, 243–49; William Kingdon Clifford, *Lectures and Essays,* ed. Frederick Pollock (London: Macmillan, 1879), 2:210; Mill, *Logic,* p. 410; R. H. Hutton, "'The Metaphysical Society,' A Reminiscence," *Nineteenth Century,* August 1885, pp. 191–92. Hutton's recreation of this Victorian symposium is fictional. Powell, "On the Study of the Evidences of Christianity," *Essays and Reviews,* 12th ed. (London: Longmans, Green, 1869), p. 169.

9. *DLL,* 2:365; *MLD,* 2:443; Stecher, *Letters,* p. 205.

10. Gray, *Science and Religion,* pp. 94–95.

11. Tyndall, *Fragments of Science,* 2:136–37; Huxley, *Science and Christian Tradition,* p. 70; LeConte, quoted in Gray, *Darwiniana,* p. 217; Nägeli, "Limits of Natural Knowledge," p. 561; *Descent,* p. 147; Powell, *Essays,* p. 355; George John Romanes [Physicus], *A Candid Examination of Theism* (Boston: Houghton, Osgood, 1878), pp. 64–65; John William Dawson, *The Story of the Earth and Man* (New York: Harper, 1873), pp. 339–43; Hull, *Darwin and His Critics,* pp. 268–69.

The conflict between the positivists and those who would allegedly overthrow "the constancy of the well-ascertained Laws of Nature" is graphically presented in the confrontation between A. R. Wallace and W. B. Carpenter over spiritualism. See Wallace, *On Miracles and Modern Spiritualism,* and W. B. Carpenter, *Mesmerism, Spiritualism, etc. Historically & Scientifically Considered* (New York: D. Appleton, 1887). The quotation is from Carpenter, p. vii.

12. Powell, "Evidences," pp. 129–30.

13. Foucault, *Order,* p. xv. It is noteworthy that latter-day "creationists" not only reject

modern evolutionary theory but also positivism as a scientific episteme.

14. The devotion to the pursuit of "pure truth" was the spiritual dimension of positive science. What communion with the mind of God was to the science of the old episteme, truth for its own sake was to the new. In a universe in which God was redundant, truth was the reigning deity.

15. For an illustration of how a theist could have his religious doctrines trimmed by participation in the positive episteme, see Peter J. Bowler, "Edward Drinker Cope and the Changing Structure of Evolutionary Theory," *Isis,* 68 (1977): 264.

16. Kuhn, *Scientific Revolutions,* pp. 94–95.

17. Oscar Schmidt, *The Doctrine of Descent and Darwinism* (New York: D. Appleton, 1880), p. 83.

18. Foucault, *Order,* p. 217.

19. *DLL,* 1:519.

20. Foucault, *Order,* p. 269; Richard W. Burkhardt, *The Spirit of System: Lamarck and Evolutionary Biology* (Cambridge: Harvard University Press, 1977), pp. 46–47.

21. For Darwin's interest in laymen as readers and as proponents of evolution, see *DLL,* 1:509; 2:39. For his doubts about the wisdom of publishing his orchid book "in a semi-popular form," however, see ibid., 2:445.

22. Huxley, *Lay Sermons,* p. 283. See Ruse, "Two Biological Revolutions," p. 37; Peter J. Bowler, "Hugo DeVries and Thomas Hunt Morgan: The Mutation Theory and the Spirit of Darwinism," *Annals of Science,* 35 (1978): 55–73.

Bibliography

Primary Sources

Agassiz, Louis. *Essay on Classification.* Edited by Edward Lurie. Cambridge: Harvard University Press, 1962.

———. "The Natural Relations between Animals and the Elements in Which They Live." *American Journal of Science,* 2d ser., 9 (1850): 369–94.

———. "The Primitive Diversity and Number of Animals in Geological Times." *American Journal of Science,* 2d ser., 17 (1854): 309–24.

———. "Professor Agassiz on the Origin of Species." *American Journal of Science,* 2d ser., 30 (1860): 142–55.

———. *The Structure of Animal Life.* 2d ed. New York: Scribner, 1866.

Agassiz, Louis, and Agassiz, Elizabeth C. *A Journey in Brazil.* Boston: Houghton Mifflin, 1895.

Agassiz, Louis, and Gould, Augustus A. *Principles of Zoology.* Boston: Gould, Kendall and Lincoln, 1848.

American Association for the Advancement of Science, *Proceedings* 3 (1850): 71, 73.

Anon. "The Immutability of Nature." *Quarterly Review* 110 (1861): 368–400.

———. "M. Comte's Course of Positive Philosophy." *Edinburgh Review* [American edition] 67 (1838): 143–63.

———. Review of *Explanations: Sequel to the Vestiges of Creation. American Journal of Science,* 2d ser., 1 (1846): 250–54.

———. Review of *Introduction to the Study of Foraminifera* by William B. Carpenter. *Athenaeum* 28 March 1863, pp. 417–19.

———. Review of *The Wonders of Geology* by Gideon A. Mantell. *American Journal of Science* 39 (1840): 1–18.

Argyll, Duke of. *Autobiography and Memoirs.* Edited by the Dowager Duchess of Argyll. 2 vols. London: John Murray, 1906.

———. "A Fourth State of Matter." *Nature,* 21 (1880): 168.

———. *Geology: Its Past and Present.* Glasgow: Griffin, 1859.

———. "A Great Lesson." *Nineteenth Century,* September 1887, pp. 293–309.

———. *Inaugural Address . . . on . . . Installation as Chancellor of the University of St. Andrews.* Edinburgh: Stillie, 1852.

———. "Mr. Herbert Spencer and Lord Salisbury on Evolution." *Nineteenth Century,* March–April 1897, pp. 387–404, 569–87.

———. "On Mr. Wallace's Theory of Birds' Nests." *Journal of Travel and Natural History* 1 (1868–69): 276–87.
———. *Organic Evolution Cross-examined or, Some Suggestions on the Great Secret of Biology.* London: John Murray, 1898.
———. *The Philosophy of Belief, or Law in Christian Theology.* London: John Murray, 1896.
———. "The Power of Loose Analogies." *Nineteenth Century,* December 1887, pp. 745–65.
———. *Primeval Man: An Examination of Some Recent Speculations.* New York: J. W. Lovell, n.d.
———. "Professor Huxley on Canon Liddon." *Nineteenth Century,* March 1887, pp. 321–29.
———. *The Reign of Law.* 5th ed. New York: Lovell, n.d.
———. "Science, Falsely So-Called." *Nineteenth Century,* May 1887, pp. 771–74.
———. *The Unity of Nature.* New York: G. P. Putnam's Sons, 1885.
———. *What Is Science?* Edinburgh: David Douglas, 1898.
Aveling, Edward. *The Religious Views of Charles Darwin.* London: Free Thought, 1883.
Bell, Charles. *The Hand.* London: Pickering, 1834.
Buckland, William. *Geology and Mineralogy Considered with Reference to Natural Theology.* 2 vols. London: Pickering, 1836.
Carpenter, William B. *Mesmerism, Spiritualism etc. Historically & Scientifically Considered.* New York: D. Appleton, 1887.
———. *Nature and Man.* London: Kegan, Paul, 1888.
[Chambers, Robert]. *Vestiges of the Natural History of Creation.* London: Churchill, 1844.
[———]. *Vestiges of the Natural History of Creation with a Sequel.* New York: Colyer, 1846.
Clifford, William Kingdon. *Lectures and Essays by the Late William Kingdon Clifford.* Edited by Frederick Pollock. 2 vols. London: Macmillan, 1879.
Colenso, John William. *The Pentateuch and Book of Joshua Critically Examined.* 2d ed. London: Longmans, Green, 1862.
[Cope, Edward Drinker]. Editorial. *American Naturalist* 20 (1886): 708–9.
———. "The Foundations of Theism." *Monist* 3 (1893): 623–29.
———. *Theology of Evolution.* Philadelphia: Arnold, 1887.
Dana, James Dwight. "Agassiz's Contributions to the Natural History of the United States." *American Journal of Science,* 2d ser., 25 (1858): 202–16; 321–41.
———. "On American Geological History." *American Journal of Science,* 2d ser., 22 (1856): 305–34.
———. *Manual of Geology.* 3d ed. New York: Ivison, 1880.
———. "Presidential Address." *American Association for the Advancement of Science, Proceedings* 9 (1854): 1–6.
———. "Science and the Bible." *Bibliotheca Sacra* 13 (1856): 80–129.

———. "Thoughts on Species." *American Journal of Science,* 2d ser., 24 (1857): 305–16.

Darwin Charles. *The Autobiography of Charles Darwin, 1809–1882.* Edited by Nora Barlow. New York: Harcourt, Brace, 1958.

———. *Charles Darwin's Diary of the Voyage of H. M. S. "Beagle."* Edited by Nora Barlow. Cambridge: Cambridge University Press, 1934.

———. *Charles Darwin's Natural Selection.* Edited by R. C. Stauffer. Cambridge: Cambridge University Press, 1975.

———. *The Collected Papers of Charles Darwin.* Edited by Paul H. Barrett. 2 vols. Chicago: University of Chicago Press, 1977.

———. *Darwin and Henslow: The Growth of an Idea, Letters 1831–1860.* Edited by Nora Barlow. Berkeley: University of California Press, 1967.

———. "Darwin's Early and Unpublished Notebooks." In *Darwin on Man,* edited by Howard E. Gruber and Paul H. Barrett. New York: Dutton, 1974.

———. "Darwin's Notebooks on Transmutation of Species." Edited by Gavin De Beer. *Bulletin of the British Museum (Natural History) Historical Series* 2, nos. 2–6 (1960–61).

———. "Darwin's Notebooks on Transmutation of Species, Part VI." Edited by Gavin De Beer, M. J. Rowlands, and B. M. Skramovsky. *Bulletin of the British Museum (Natural History) Historical Series* 3, no. 5 (1967).

———. "Darwin's Ornithological Notes." Edited by Nora Barlow. *Bulletin of the British Museum (Natural History) Historical Series* 2, no. 7 (1963).

———. *The Descent of Man and Selection in Relation to Sex.* 2d ed. New York: D. Appleton, 1897.

———. *The Effects of Cross and Self Fertilisation in the Vegetable Kingdom.* New York: D. Appleton, 1895.

———. *The Expression of the Emotions in Man and Animals.* New York: D. Appleton, 1898.

———. *The Foundations of the Origin of Species: Two Essays Written in 1842 and 1844.* Edited by Francis Darwin. Cambridge: Cambridge University Press, 1909.

———. "Geology." In *A Manual of Scientific Enquiry,* edited by John F. W. Herschel. 2d ed. London: John Murray, 1851.

———. *Journal and Remarks, 1832–1836.* Narrative of the Surveying Voyages of His Majesty's Ships Adventure and Beagle . . . , vol. 3. London: Colburn, 1839.

———. *Journal of Researches . . .* London: John Murray, 1897.

———. *The Life and Letters of Charles Darwin.* Edited by Francis Darwin. 2 vols. New York: D. Appleton, 1896.

———. *More Letters of Charles Darwin.* Edited by Francis Darwin. 2 vols. New York: D. Appleton, 1903.

———. *The Origin of Species, by Charles Darwin: A Variorum Text.* Edited by Morse Peckham. Philadelphia: University of Pennsylvania Press, 1959.

———. "Some Unpublished Letters of Charles Darwin." Royal Society of London. *Notes and Records* 14 (1959): 12–60.

———. *The Structure and Distribution of Coral Reefs.* London: Smith, Elder, 1842.
———. *The Variation of Animals and Plants under Domestication.* 2d ed. 2 vols. New York: D. Appleton, 1896.
———. *On the Various Contrivances by which British and Foreign Orchids are Fertilised by Insects and on the Good Effects of Inter-crossing.* London: John Murray, 1862.
Darwin, Emma. *Emma Darwin: A Century of Family Letters, 1796–1896.* Edited by Henrietta Litchfield. 2 vols. New York: D. Appleton, 1915.
Dawson, John William. *The Story of the Earth and Man.* New York: Harper, 1873.
Du Bois-Reymond, Emile. "The Limits of Our Knowledge of Nature." *Popular Science Monthly,* May 1874, pp. 17–32.
Evans, Sabastian. "Sir John Lubbock on the Origin of Civilisation." *Nature,* 3 (1870–71): 362–65.
Fiske, John. "Draper's Science and Religion." *Nation,* 25 November 1875, pp. 343–45.
Foster, J. W., and Whitney, J. D. "On the Azoic System, as Developed in the Lake Superior Land District." *American Association for the Advancement of Science, Proceedings* 5 (1851): 4–7.
Foulerton, J. "The Duke of Argyll on the Unity of Nature." *Journal of Science,* 3d ser., 3 (1880): 671–79.
Goodwin, Henry, Bishop of Carlisle. "The Uniformity of Nature." *Popular Science Monthly,* December 1885, pp. 248–60.
Graham, William. *The Creed of Science: Religious, Moral, and Social.* London: Kegan Paul, 1881.
Gray, Asa. *Darwiniana.* Edited by A. Hunter Dupree. Cambridge: Harvard University Press, 1963.
[———]. "Introductory Essay, in Dr. Hooker's Flora of New Zealand." *American Journal of Science,* 2d ser., 17 (1854): 241–52, 334–50.
———. *Letters of Asa Gray.* Edited by Jane Loring Gray. 2 vols. Boston: Houghton, Mifflin, 1893.
———. *Natural Science and Religion.* New York: Scribners, 1880.
———. "Natural Selection and Natural Theology." *Nature,* 27 (1882–83): 291–92, 527–28.
———. Review of *On the Various Contrivances by which . . . Orchids are Fertilised . . .* by Charles Darwin. *American Journal of Science,* 2d ser., 34 (1862): 138–44.
[———]. Review of *The Variation of Animals and Plants under Domestication* by Charles Darwin. *Nation,* 19 March 1868, pp. 234–36.
Harvey, William H. "Darwin on the Origin of Species." *Gardener's Chronicle and Agricultural Gazette,* 18 February 1860, pp. 145–46.
Herschel, John F. W. *A Preliminary Discourse on the Study of Natural Philosophy.* London: Longman, 1831.
———. *Physical Geography.* Edinburgh: Black, 1861.
Hitchcock, Edward. *Elementary Geology.* 13th ed. New York: Ivison and Phinney, 1859.

———. *The Religion of Geology and Its Connected Sciences.* New ed. Boston: Phillips, Sampson, 1859.

Hodge, Charles. *What is Darwinism?* New York: Charles Scribner, 1874.

Hooker, Joseph Dalton. *Flora Novae Zealandicae* and *Flora Tasmaniae.* The Botany of the Antarctic Voyage of H. M. Discovery Ships "Erebus" and "Terror" in the Years 1839–1843, vols. 2 and 3. London: Reeve, 1847–60.

———. *Life and Letters of Sir Joseph Dalton Hooker.* Edited by Leonard Huxley. 2 vols. London: John Murray, 1918.

———. Presidential Address, British Association for the Advancement of Science. *The Times* (London), 20 August 1868.

Hopkins, William. "Anniversary Address of the President." *Journal of the Geological Society of London* 8 (1852): xxi–lxxx.

Hutton, F. W. "Some Remarks on Mr. Darwin's Theory." *Geologist* 4 (1861): 132–36, 183–88.

Hutton, R. H. "'The Metaphysical Society.' A Reminiscence." *Nineteenth Century,* August 1885, pp. 177–96.

Huxley, Thomas Henry. *Darwiniana.* New York: D. Appleton, 1898.

———. *Lay Sermons, Addresses and Reviews.* New York: D. Appleton, 1871.

———. *Life and Letters of Thomas Henry Huxley.* Edited by Leonard Huxley. 2 vols. New York: D. Appleton, 1900.

———. *Methods and Results.* New York: D. Appleton, 1898.

———. "On the Persistent Types of Animal Life." *Scientific Memoirs of Thomas Henry Huxley.* 4 vols. London: Macmillan, 1898–1902.

———. Review of *The Origin of Species* by Charles Darwin. *The Times* (London), 26 December 1859.

———. *Science and Christian Tradition.* New York: D. Appleton, 1898.

———. *Science and Hebrew Tradition.* New York: D. Appleton, 1896.

———. "Science and Pseudo-Science." *Nineteenth Century,* April 1887, pp. 481–98.

———. "Scientific and Pseudo-Scientific Realism." *Nineteenth Century,* February 1887, pp. 191–205.

———. "Science and Religion," *The Builder* 18 (1859): 35–36.

Jevons, E. Stanley. "Evolution and the Doctrine of Design." *Popular Science Monthly,* May 1874, pp. 98–100.

Kirby, William. *On the Power, Wisdom and Goodness of God as Manifested in the Creation of Animals and in their History, Habits and Instincts.* 2 vols. London: W. Pickering, 1835.

LeConte, Joseph. *Evolution and Its Relation to Religious Thought.* 2d ed. New York: D. Appleton, 1891.

———. "Lectures on Coal." *Annual Report of the Smithsonian Institution, 1857.* Washington: Smithsonian Institution, 1858.

Lyell, Charles. "Anniversary Address of the President." *Journal of the Geological Society of London* 6 (1850): xxvii–lxvi.

———. "Anniversary Address of the President." *Journal of the Geological Society of*

London 7 (1851): xxv–lxxvi.

———. *The Geological Evidence of the Antiquity of Man with Remarks on Theories of the Origin of Species by Variation.* 2d American ed. Philadelphia: Childe, 1863.

———. *Life, Letters and Journals of Sir Charles Lyell, Bart.* Edited by Mrs. Lyell. 2 vols. London: John Murray, 1881.

———. *Principles of Geology.* 3 vols. London: John Murray, 1830–33.

———. *Principles of Geology.* 8th ed. London: John Murray, 1850.

———. *Principles of Geology.* 11th ed. 2 vols. New York: D. Appleton, 1873.

———. Review of *Transactions of the Geological Society of London, 1824. Quarterly Review* 34 (1826): 507–40.

———. *Sir Charles Lyell's Scientific Journals on the Species Question.* Edited by Leonard G. Wilson. New Haven: Yale University Press, 1970.

Mill, John Stuart. *A System of Logic.* 8th ed. London: Longmans, 1961.

Miller, Hugh. *Footprints of the Creator, or the Asterolepsis of Stromness.* 11th ed. Edinburgh: Nimmo, 1869.

Mivart, St. George Jackson. *On the Genesis of Species.* London: Macmillan, 1871.

Murchison, Roderick I.; Verneuil, Eduard de; and Keyserling, Alexander von. *The Geology of Russia in Europe and the Ural Mountains.* 2 vols. London: John Murray, 1845.

Murray, Andrew. "On the Disguises of Nature: Being an Inquiry into the Laws Which Regulate External Forms and Colour in Plants and Animals." *Edinburgh New Philosophical Journal* 68 (1860): 66–90.

Nägeli, Carl von. "The Limits of Natural Knowledge." *Nature,* 16 (1877): 531–35, 559–63.

Nature, 25 (1881–82): 7; 37 (1887–88): 25–26, 53–54, 246, 272, 293, 317–18, 342, 363–64; 59 (1898–99): 217–19, 246–47, 269, 317–18.

Newton, Alfred. "The Early Days of Darwinism." *Macmillan's Magazine,* February 1888, pp. 241–49.

Owen, Richard. *Address.* London: n. p., 1858.

———. *On the Anatomy of Vertebrates.* 3 vols. London: Longmans, Green, 1866–68.

———. *On the Archetype and Homologies of the Vertebrate Skeleton.* London: Van Voorst, 1848.

———. "On the Argument of 'Infirmity' in Mr. Lewes' Review of the Reign of Law." *Fraser's Magazine,* October 1867, pp. 531–33.

———. *On the Classification and Geographical Distribution of the Mammalia. . . .* London: Parker, 1859.

———. *On the Nature of Limbs.* London: Van Voorst, 1849.

———. *Palaeontology, or a Systematic Summary of Extinct Animals and their Geological Relations.* 2d ed. Edinburgh: Black, 1861.

———. "The Principal Forms of the Skeleton." *Orr's Circle of the Sciences: Organic Nature.* 4 vols. London: Orr, 1854.

Paley, William. *Natural Theology* in *Works of William Paley, D. D.* London: Tegg, 1853.

Pengelly, Hester, ed. *A Memoir of William Pengelly, of Torquay, F. R. S., Geologist, with a Selection from His Correspondence.* London: John Murray, 1897.

Phillips, John. "Anniversary Address of the President." *Journal of the Geological Society of London* 16 (1860): xxvii–lv.

———. *Life on Earth: Its Origin and Succession.* London: Macmillan, 1860.

Portlock, Major General Joseph E. "Anniversary Address of the President." *Journal of the Geological Society of London* 14 (1858): xxiv–clxiii.

Powell, Baden. *Essays on the Spirit of the Inductive Philosophy, the Unity of Worlds and the Philosophy of Creation.* London: Longmans, 1855.

———. "On the Study of the Evidences of Christianity." In *Essays and Reviews.* 12th ed. London: Longmans, Green, 1869.

Prout, William. *Chemistry, Meteorology and the Function of Digestion Considered with Reference to Natural Theology.* London: W. Pickering, 1834.

Pusey, Edward B. "Dr. Pusey on Science and Theology." *The Guardian,* 20 November 1878.

———. *Un-Science, Not Science, Adverse to Faith.* 2 ed. London: Parker, 1878.

Roget, Peter Mark. *Animal and Vegetable Physiology.* 2 vols. London: W. Pickering, 1834.

[Romanes, George John]. *A Candid Examination of Theism.* Boston: Houghton, Osgood, 1878.

———. "Natural Selection and Natural Theology." *Contemporary Review* 42 (1882): 536–43.

———. "Natural Selection and Natural Theology." *Nature,* 27 (1882–83): 362–64, 528–29.

Schiller, F. C. S. "Darwinism and Design." *Contemporary Review* 71 (1897): 867–83.

Schmidt, Oscar. *The Doctrine of Descent and Darwinism.* New York: D. Appleton, 1880.

Sedgwick, Adam. *Discourse on the Studies of the University of Cambridge.* 5th ed. London: Parker, 1850.

———. *The Life and Letters of the Reverend Adam Sedgwick.* Edited by John Willis Clark and T. McHughes. 2 vols. Cambridge: Cambridge University Press, 1890.

Spencer, Herbert. *An Autobiography.* 2 vols. New York: D. Appleton, 1904.

———. "The Development Hypothesis." In *Essays, Scientific, Political, and Speculative.* 3 vols. New York: D. Appleton, 1896.

———. *The Principles of Sociology.* 5 vols. New York: D. Appleton, 1896.

Stecher, Robert M. "The Darwin-Innes Letters: The Correspondence of an Evolutionist with His Vicar, 1848–1884." *Annals of Science* 17 (1961): 201–58.

The Times (London), 30 September 1844; 15 September 1864.

Tylor, Edward Burnett. *Primitive Culture.* 2 vols. London: John Murray, 1871.

Tyndall, John. *Fragments of Science.* 2 vols. New York: D. Appleton, 1897.

Wallace, Alfred Russel. *On Miracles and Modern Spiritualism. Three Essays.* 2d ed. London: Trübner, 1881.

———. *Natural Selection and Tropical Nature.* Rev. ed. London: Macmillan, 1891.

Whewell, William. *Astronomy and General Physics Considered with Reference to Natural Theology.* London: Pickering, 1836.

———. *History of the Inductive Sciences.* 3d ed. 3 vols. London: Parker, 1857.

[———]. Review of *Principles of Geology,* vol. 2, by Charles Lyell. *Quarterly Review* 47 (1832): 103–32.

———. "On the Wave of Translation in Connexion with the Northern Drift." *Journal of the Geological Society of London* 3 (1847): 227–32.

Wollaston, T. Vernon. *On the Variation of Species, with Especial Reference to the Insecta; Followed by an Inquiry into the Nature of Genera.* London: Van Voorst, 1856.

Works Progress Administration. *Calendar of the Letters of Charles Robert Darwin to Asa Gray.* Boston: Historical Records Survey, 1939.

[Youmans, W. J.]. "The Duke of Argyll on Evolution," *Popular Science Monthly,* July 1897, pp. 411–15.

Secondary Sources

Abbagnano, Nicola. In *Encyclopedia of Philosophy,* s.v. "positivism."

Annan, Noel G. *The Curious Strength of Positivism in English Political Thought.* Oxford: Oxford University Press, 1959.

Badash, Lawrence. "The Completeness of Nineteenth-Century Science." *Isis* 63 (1972): 48–58.

Barrett, Paul H. "The Sedgwick-Darwin Geologic Tour of North Wales." *Proceedings of the American Philosophical Society* 118 (1974): 146–64.

Bartholomew, Michael. "Huxley's Defence of Darwin." *Annals of Science* 23 (1975): 525–35.

———. "Lyell and Evolution: An Account of Lyell's Response to the Prospect of an Evolutionary Ancestry for Man." *British Journal for the History of Science* 6 (1973): 261–303.

———. "The Non-progress of Non-progression: Two Responses to Lyell's Doctrine." *British Journal for the History of Science* 9 (1976): 166–74.

Basalla, George et al., eds. *Victorian Science.* Garden City, N.Y.: Doubleday, 1970.

Bowler, Peter J. "Darwinism and the Argument from Design: Suggestions for a Reevaluation." *Journal of the History of Biology* 10 (1977): 29–44.

———. "Darwin's Concepts of Variation." *Journal of the History of Medicine* 29 (1974): 196–212.

———. "Edward Drinker Cope and the Changing Structure of Evolutionary Theory." *Isis* 68 (1977): 249–65.

———. "Evolution in the Enlightenment." *History of Science* 12 (1974): 159–83.

———. *Fossils and Progress.* New York: Science History Publications, 1976.

———. "Hugo De Vries and Thomas Hunt Morgan: The Mutation Theory and the Spirit of Darwinism." *Annals of Science* 35 (1978): 55–73.

Brown, Alan. *The Metaphysical Society: Victorian Minds in Conflict, 1869–1880.* New York: Columbia University Press, 1947.

Burchfield, Joe D. *Lord Kelvin and the Age of the Earth.* New York: Science History Publications, 1975.

Burckhardt, Richard W., Jr. *The Spirit of System: Lamarck and Evolutionary Biology.*

Cambridge: Harvard University Press, 1977.

Campbell, John Angus. "Nature, Religion, and Emotional Response: A Reconsideration of Darwin's Affective Decline." *Victorian Studies* 18 (1974): 159–74.

Cannon, W. F. "The Basis of Darwin's Achievement: A Revaluation." *Victorian Studies* 5 (1961): 109–34.

———. "Charles Lyell Is Permitted to Speak for Himself: An Abstract." In *Towards a History of Geology,* edited by Cecil J. Schneer. Cambridge: M. I. T. Press, 1969.

———. "Charles Lyell, Radical Actualism, and Theory." *British Journal for the History of Science* 9 (1976): 104–20.

———. "Darwin's Vision in *On the Origin of Species.*" In *Art of Victorian Prose,* edited by George Levine and William Madden. New York: Oxford University Press, 1968.

———. "The Impact of Uniformitarianism: Two Letters from John Herschel to Charles Lyell, 1836–1837." *Proceedings of the American Philosophical Society* 105 (1961): 301–14.

———. "John Herschel and the Idea of Science." *Journal of the History of Ideas* 20 (1961): 215–39.

———. "The Problem of Miracles in the 1830's." *Victorian Studies* 4 (1960): 5–32.

Cantor, Geoffrey. "The Reception of the Wave Theory of Light in Britain: A Case Study Illustrating the Role of Methodology in Scientific Debate." *Historical Studies in the Physical Sciences* 6 (1975): 109–32.

Cassirer, Ernst. *The Problem of Knowledge: Philosophy, Science and History since Hegel.* New Haven: Yale University Press, 1950.

Chadwick, Owen. *The Secularization of the European Mind in the Nineteenth Century.* Cambridge: Cambridge University Press, 1975.

Charlton, D. G. *Positive Thought in France during the Second Empire.* Oxford: Clarendon Press, 1959.

Colp, Ralph, Jr. *To Be an Invalid: The Illness of Charles Darwin.* Chicago: University of Chicago Press, 1977.

Crombie, A. C. "Darwin's Scientific Method." *Actes du IXe congres, Academie internationale d'histoire des sciences* 1 (1960): 324–62.

Davies, Gordon L. *The Earth in Decay: A History of British Geomorphology, 1578–1878.* New York: Science History Publications, 1969.

De Beer, Gavin. *Charles Darwin: A Scientific Biography.* Garden City, N.Y.: Doubleday, 1964.

Dupree, A. Hunter. *Asa Gray, 1810–1888.* New York: Atheneum, 1968.

Egerton, Frank. "Darwin's Method or Methods?" *Studies in History and Philosophy of Science* 2 (1971): 281–86.

———. "Humboldt, Darwin and Population." *Journal of the History of Biology* 3 (1970): 325–60.

Ellegård, Alvar. *Darwin and the General Reader.* Göteborg: University of Gothenburg, 1958.

Farley, John. *The Spontaneous Generation Controversy from Descartes to Oparin.* Baltimore: Johns Hopkins University Press, 1977.

Fay, Margaret A. "Did Marx Offer to Dedicate *Capital* to Darwin? *Journal of the History of Ideas* 39 (1978): 133–46.

Feuer, Lewis S. "Is the 'Darwin-Marx Correspondence' Authentic?" *Annals of Science* 32 (1975): 1–12.

Feuer, Lewis, et al. "On the Darwin-Marx Correspondence." *Annals of Science* 33 (1976): 383–94.

Figlio, Karl M. "The Metaphor of Organization: An Historiographical Perspective on the Bio-medical Sciences of the Early Nineteenth Century." *History of Science* 14 (1976): 17–53.

Fleming, Donald. "Charles Darwin, the Anaesthetic Man." *Victorian Studies* 4 (1961): 219–36.

Foucault, Michel. *L'archéologie du savoir.* Paris: Gallimard, 1969. Translated by A. M. Sheridan Smith as *The Archaeology of Knowledge.* New York: Harper and Row, 1972.

———. *Les mots et les choses: une archéologie des sciences humaines.* Paris: Gallimard, 1966. Translated as *The Order of Things, An Archaeology of the Human Sciences.* London: Tavistock, 1970.

Geison, Gerald L. "Darwin and Heredity: The Evolution of His Hypothesis of Pangenesis." *Journal of the History of Medicine* 24 (1969): 375–411.

George, T. Neville. "Charles Lyell: The Present Is Key to the Past." *Philosophical Journal: Transactions of the Royal Philosophical Society of Glasgow* 13 (1976): 3–24.

Ghiselin, Michael T. "The Individual in the Darwinian Revolution." *New Literary History* 3 (1971): 113–34.

———. "Mr. Darwin's Critics, Old and New." *Journal of the History of Biology* 6 (1973): 155–65.

———. *The Triumph of the Darwinian Method.* Berkeley: University of California Press, 1969.

Gillespie, Neal C. "The Duke of Argyll, Evolutionary Anthropology, and the Art of Scientific Controversy." *Isis* 68 (1977): 40–54.

Gillispie, Charles C. *The Edge of Objectivity.* Princeton: Princeton University Press, 1960.

Gould, Stephen Jay. *Ontogeny and Phylogeny.* Cambridge: Harvard University Press, 1977.

Greene, John C. "Darwin as a Social Evolutionist." *Journal of the History of Biology* 10 (1977): 1–28.

———. "The Kuhnian Paradigm and the Darwinian Revolution." In *Perspectives in the History of Science and Technology,* edited by Duane H. D. Roller. Norman: University of Oklahoma Press, 1971.

———. "Reflections on the Progress of Darwin Studies." *Journal of the History of Biology* 8 (1975): 243–74.

Grinnell, George. "The Rise and Fall of Darwin's First Theory of Transmutation." *Journal of the History of Biology* 7 (1974): 259–73.

Gruber, Howard E., and Barrett, Paul H. *Darwin on Man.* New York: Dutton, 1974.

Gruber, Howard E., and Gruber, V. "The Eye of Reason: Darwin's Development during the *Beagle* Voyage." *Isis* 53 (1962): 186–200.

Gruber, Jacob W. *A Conscience in Conflict: The Life of St. George Jackson Mivart.* New York: Columbia University Press, 1960.

Guralnick, Stanley M. "Geology and Religion before Darwin: The Case of Edward Hitchcock, Theologian and Geologist (1793–1864)." *Isis* 63 (1972): 529–43.

———. "Sources of Misconception on the Role of Science in the Nineteenth-Century American College." *Isis* 65 (1974): 352–66.

Herbert, Sandra. "The Place of Man in the Development of Darwin's Theory of Transmutation. Part 1 to July, 1837." *Journal of the History of Biology* 7 (1974): 216–58.

———. "The Place of Man in the Development of Darwin's Theory of Transmutation. Part II." *Journal of the History of Biology* 10 (1977): 155–227.

Hooykaas, R. *The Principle of Uniformity in Geology, Biology and Theology.* Leiden: Brill, 1963.

Hull, David L. "Charles Darwin and Nineteenth-Century Philosophies of Science." In *Foundations of Scientific Method: The Nineteenth Century,* edited by N. Giere and R. S. Westfall. Bloomington: Indiana University Press, 1973.

———. "Metaphysics of Evolution." *British Journal for the History of Science* 3 (1967): 309–37.

———. *Philosophy of Biological Science.* Englewood Cliffs, N.J.: Prentice-Hall, 1974.

———, ed. *Darwin and His Critics: The Reception of Darwin's Theory of Evolution by the Scientific Community.* Cambridge: Harvard University Press, 1973.

Kaplan, Abraham. *The Conduct of Inquiry: Methodology for Behavioral Science.* Scranton, Pa.: Chandler, 1964.

Kavaloski, Vincent C. "The *Vera Causa* Principle: An Historico-Philosophical Study of a Metatheoretical Concept from Newton through Darwin." Ph.D. dissertation, University of Chicago, 1974.

Kohn, E. David. "Charles Darwin's Path to Natural Selection." Ph.D. dissertation, University of Massachusetts, 1975.

Kolakowski, Leszek. *The Alienation of Reason: A History of Positivist Thought.* Garden City, N.Y.: Doubleday, 1968.

Kottler, Malcolm Jay. "Alfred Russel Wallace, the Origin of Man, and Spiritualism." *Isis* 65 (1974): 145–92.

Kuhn, Thomas S. "Second Thoughts on Paradigms." In *The Structure of Scientific Theories,* edited by Frederick Suppe. Urbana: University of Illinois Press, 1974.

———. *The Structure of Scientific Revolutions.* 2d ed. Chicago: University of Chicago Press, 1970.

Lakatos, Imre, and Musgrave, Alan, eds. *Criticism and the Growth of Knowledge.* Cambridge: Cambridge University Press, 1970.

Leary, David. "Michel Foucault, an Historian of the *Sciences Humaines.*" *Journal of the History of the Behavioral Sciences* 12 (1976): 286–93.

Limoges, Camille. *La Sélection naturelle. Étude sur la première constitution d'un concept*

(1837–1859). Paris: Presses universitaires de France, 1970.

Lurie, Edward. *Louis Agassiz. A Life in Science.* Chicago: University of Chicago Press, 1960.

McElligott, John F. "Before Darwin: Religion and Science as Presented in American Magazines, 1830–1860." Ph.D. dissertation, New York University, 1973.

MacLeod, Roy M. "Evolutionism and Richard Owen, 1830–1868: An Episode in Darwin's Century." *Isis* 56 (1965): 259–80.

Mandelbaum, Maurice. "Darwin's Religious Views." *Journal of the History of Ideas* 19 (1958): 363–78.

———. *History, Man and Reason: A Study of Nineteenth Century Thought.* Baltimore: Johns Hopkins University Press, 1971.

Manier, Edward. *The Young Darwin and His Cultural Circle.* Dordrecht, Holland: D. Reidel, 1978.

Mayr, Ernst. *Evolution and the Diversity of Life: Selected Essays.* Cambridge: Harvard University Press, 1976.

———. Introduction to *On the Origin of Species,* by Charles Darwin. Cambridge: Harvard University Press, 1964.

———. "Open Problems of Darwin Research." *Studies in History and Philosophy of Science* 2 (1971): 273–80.

Millhauser, Milton. *Just before Darwin: Robert Chambers and 'Vestiges.'* Middletown, Conn.: Wesleyan University Press, 1959.

Numbers, Ronald L. *Creation by Natural Law.* Seattle: University of Washington Press, 1977.

Osborn, Henry Fairfield. *Cope: Master Naturalist.* Princeton: Princeton University Press, 1931.

Ospovat, Dov. "The Influence of Karl Ernst von Baer's Embryology, 1828–1859: A Reappraisal in Light of Richard Owen's and William B. Carpenter's Palaeontological Application of 'Von Baer's Law.'" *Journal of the History of Biology* 10 (1976): 1–28.

Page, Leroy E. "Diluvialism and Its Critics in Great Britain in the Early Nineteenth Century." In *Towards a History of Geology,* edited by Cecil J. Schneer. Cambridge: M. I. T. Press, 1969.

———. "The Rivalry between Charles Lyell and Roderick Murchison." *British Journal for the History of Science* 9 (1976): 156–65.

Phillips, Derek L. "Paradigms and Incommensurability." *Theory and Society* 2 (1975): 37–61.

Porter, Roy. "Charles Lyell and the Principles of the History of Geology." *British Journal for the History of Science* 9 (1976): 91–103.

Rappaport, Rhoda. "Geology and Orthodoxy: The Case of Noah's Flood in Eighteenth-Century Thought." *British Journal for the History of Science* 11 (1978): 1–18.

Rudwick, Martin J. S. "Darwin and Glen Roy: A 'Great Failure' in Scientific Method?" *Studies in History and Philosophy of Science* 5 (1974): 97–185.

———. *The Meaning of Fossils: Episodes in the History of Palaeontology.* London: MacDonald, 1972.

———. "The Strategy of Lyell's *Principles of Geology.*" *Isis* 61 (1970): 5–33.

———. "Uniformity and Progression: Reflections on the Structure of Geological Theory in the Age of Lyell," In *Perspectives in the History of Science and Technology,* edited by Duane H. D. Roller. Norman: University of Oklahoma Press, 1971.

Ruse, Michael. "Charles Darwin's Theory of Evolution: An Analysis." *Journal of the History of Biology* 8 (1975): 219–42.

———. "Charles Lyell and the Philosophers of Science." *British Journal for the History of Science* 9 (1976): 121–31.

———. "The Darwin Industry." *History of Science* 12 (1974): 43–58.

———. "Darwin's Debt to Philosophy: An Examination of the Influence of the Philosophical Ideas of John F. W. Herschel and William Whewell on the Development of Charles Darwin's Theory of Evolution." *Studies in History and Philosophy of Science* 6 (1975): 159–81.

———. "The Nature of Scientific Models: Formal v. Material Analogy." *Philosophy of the Social Sciences* 3 (1973): 63–80.

———. "The Revolution in Biology." *Theoria* 35 (1971): 17–38.

———. "Two Biological Revolutions." *Dialectica* 25 (1971): 17–38.

———. "The Value of Analogical Models in Science." *Dialogue* 12 (1973): 246–53.

Scheffler, Israel. *Science and Subjectivity.* Indianapolis: Bobbs-Merrill, 1967.

Schweber, Silvan S. "The Origin of the *Origin* Revisited." *Journal of the History of Biology* 10 (1977): 229–316.

Shaffer, E. S. "The Archaeology of Michel Foucault." *Studies in History and Philosophy of Science* 7 (1976): 269–75.

Shapere, Dudley, "The Structure of Scientific Revolutions." *Philosophical Review* 73 (1964): 383–94.

Simon, W. M. *European Positivism in the Nineteenth Century.* Ithaca, N.Y.: Cornell University Press, 1963.

Staianovich, Traian. *French Historical Method: The Annales Paradigm.* Ithaca, N.Y.: Cornell University Press, 1976.

Toulmin, Stephen. *Human Understanding.* Oxford: Oxford University Press, 1972.

Trigg, Roger. *Reason and Commitment.* New York: Oxford University Press, 1973.

Turner, Frank M. *Between Science and Religion: The Reaction to Scientific Naturalism in Late Victorian England.* New Haven: Yale University Press, 1974.

Vorzimmer, Peter J. "Charles Darwin and Blending Inheritance." *Isis* 54 (1963): 371–90.

———. *Charles Darwin: The Years of Controversy: The Origin of Species and Its Critics, 1859–1882.* Philadelphia: Temple University Press, 1970.

———. "The Darwin Reading Notebooks (1838–1860)." *Journal of the History of Biology* 10 (1977): 107–52.

———. "An Early Darwin Manuscript: The 'Outline and Draft of 1839.'" *Journal of the History of Biology* 8 (1975): 191–217.

Westfall, Richard S. *Science and Religion in Seventeenth-Century England.* New Haven: Yale University Press, 1958.

Wilson, David B. "Herschel and Whewell's Version of Newtonianism." *Journal of the History of Ideas* 35 (1974): 79–97.

———. "Victorian Science and Religion." *History of Science* 15 (1977): 52–67.

Wilson, Leonard G. *Charles Lyell: The Years to 1841.* New Haven: Yale University Press, 1972.

Wilson, Major L. "Paradox Lost: Order and Progress in Evangelical Thought of Mid-Nineteenth Century America." *Church History* 44 (1975): 352–66.

Winsor, Mary P. *Starfish, Jellyfish and the Order of Life.* New Haven: Yale University Press, 1976.

Young, Robert M. "Darwin's Metaphor: Does Nature Select?" *Monist* 55 (1971): 442–503.

Index